The Single Malt
Whisky
COMPANION

A Connoisseur's Guide

The Single Malt Whisky

COMPANION

A Connoisseur's Guide

HELEN ARTHUR

PAST TIMES

A QUINTET BOOK

First published in Great Britain in 2001 by
Past Times Oxford England

Reprinted 2002

The book was designed and produced by
Quintet Publishing Ltd.
The Old Brewery
6 Blundell Street
London N7 9BH

CREATIVE DIRECTOR Richard Dewing
SENIOR EDITOR Anna Briffa
ART DIRECTOR Clare Reynolds
DESIGNER Isobel Gillan
EDITORS Anna Bennett, Olga Herrera Moya
PHOTOGRAPHERS Paul Forrester, Laura Wickenden, Colin Bowden
ILLUSTRATORS Michael Hill, Richard Chasemore

Typeset by Central Southern Typesetters, Eastbourne, UK
Manufactured in Singapore by United Graphic Pte Ltd.
Printed in China by Midas Printing Limited.

Contents

Foreword

At long last someone has taken the time and effort to dispel the myth that the world of whisky is a "man's world." There are many female writers who have written and commented on whisky, and who continue to do so quite regularly in newspapers and magazines. To the best of my knowledge, however, this is the first in-depth treatise by a woman and as such is long overdue.

This is an informative companion and a first-class guide to have at hand while tasting, sipping, and savoring your favorite whisky, or while trying out a new brand, be it on your own, or with family, or colleagues. You may not always agree with Helen's tasting notes and comments; from time to time Helen and I have an amicable difference of opinion ourselves and decide that we should return for a further tasting session. And long may such differences of opinion last since it will be a very sad day when we all agree on taste. The nuances that we enjoy are all part of the richness of the world's greatest drink.

What gives whisky its character is a question that will never be answered satisfactorily, though many theories have been put forward; the size or shape of the stills, or the geographical location of the distillery for example. We do know however that each and every whisky has contrasting pleasures, and it is left to you the imbiber to come forward with your own thoughts and opinions on the subject.

With this book Helen has given you the foundation on which to build your enjoyment of this spirit; you will become enchanted with the excitement of discovering and a whole new world of sensory experiences, which you never knew existed. Just as experts and great thinkers enjoy sharing their great thoughts so we are bound to share our thoughts upon the whiskies featured here.

While Helen hopes that, as you read this book you will appreciate not just the traditions and skills, but also capture a little of the magic that goes into whisky-making, I would urge you not to forget the tender loving care that has been passed down from father to son for generations.

Wallace Milroy, April, 1997

Author's Introduction

The word whisky conjures up a picture of an amber-colored liquid with a range of tastes and aromas. Add the word Scotch to whisky, however, and a range of other images come to mind. The amber liquid has gained a heritage, a pedigree, and a life of its own. Visions of heather-covered hills, peat bogs, cool clear water bubbling over granite rocks, the sound of bagpipes, kilted knees, stags' heads on oak-paneled walls, a roaring fire, and a cut-glass crystal decanter with goblets spring to mind.

This book is an introduction to the world of single malt whiskies, that is, whiskies which are made exclusively from malted barley and distilled at individual distilleries. There are more than 100 different whisky distilleries in Scotland so, of necessity, the bulk of this book will deal with these. Many of the fine malts they distill are never destined to be sold as a single brand but make up parts of well-known blends such as The Famous Grouse, Teacher's, and Bell's. For the pleasure of whisky connoisseurs the world over, however, a large number are bottled and sold as single malts.

The main section of this book, the Directory lists those malts which are readily accessible, either bottled by the individual distilleries or by specialists. Details are also given of rare malts, whiskies from distilleries which have ceased production and whiskies at ages and strengths not generally commercially available. The Directory also lists, in addition to the single malts distilled in Scotland, a few fine single malts produced in Ireland and Japan. Visitors are welcome at many distilleries and information of the facilities they offer and schedules are also given in the Directory section.

SCOTCH WHISKY PRODUCTION

Scotch whisky production during 1995 was equivalent to 1,407,900 70-cl bottles (approx. ¾ quart); this represents a substantial increase on 1994. Sales of single malt account for some 5 percent of this overall total.

Export figures for 1995 amounted to $30 million, some 85 percent of the total produced. Exports account for a high percentage of single malts, limited editions, and special bottlings.

WHISKY OR WHISKEY?

Generally, "whisky" is made in Scotland and "whiskey" is made in Ireland and the United States. Confusingly, however, most whiskies marketed in Japan and Canada are labeled "whisky." Malt whisky is the product of malted barley only. Grain whisky and whiskey from Ireland and the United States are produced from a wider range of cereals including rye, wheat, and corn.

AUTHOR ACKNOWLEDGMENTS

I extend grateful thanks to everyone who has helped me put this book together, for their time, hospitality, and shared knowledge. Memories of visits to distilleries, carrying out research in archives, talking to those who work in all aspects of the industry, and above all having the opportunity to taste so many fine single malts will remain with me for a very long time. Thank you.

As with all books it would have been impossible to complete this without the support of a large number of people. Firstly, of course, thanks go to Anna Briffa, my editor, for her patience and guidance. I would particularly like to thank James McEwan of Morrison Bowmore Distillers for his support and hospitality and Dr. Alan Rutherford of United Distillers for his expertise and friendship. And then thanks go to all the people in the Scotch whisky industry with whom I have worked over the years; these include Matthew Gloag of Matthew Gloag & Sons Ltd., Islay Campbell, distillery manager at Bowmore, Iain Henderson at Laphroaig, Mike Nicolson at Caol Ila, Alistair Skakles at Royal Lochnagar, Bill Bergius of Allied Distillers, Ian Urquhart at Gordon & MacPhail, Campbell Evans at the Scotch Whisky Association and Caroline Dewar. And a big thank you to Wallace Milroy, the celebrated whisky writer, and his brother John for just being there when I needed them. I also am very grateful to my team of tasters who helped me savor the wide range of single malts—Graham Cook, Sue Holmes, Charles Richardson-Bryant, Danny West and Tony Vigne.

I hope that as you read this book you will learn something new and will come to appreciate the traditions and skills that go into making whisky, and perhaps capture some of the magic as well! For without magic there would be no single malt.

Helen Arthur
Putley, Herefordshire, 1997

The Story of
Single Malt
Whisky

Whisky
Part of Scotland's Heritage

The first reference to making what we know as Scotch whisky dates from 1494 when one Friar John Cor of Lindores Abbey, near Newburgh in Fife, is cited as purchasing eight bolts of malt, which would have produced 35 cases.

Home whisky distillation, which was permitted under the law, was a logical part of a Scottish farmer's economy. During the summer, cattle were raised and barley grown to feed them in the winter months. Any surplus barley could then be used to produce a warming alcoholic beverage to counteract the cold.

Following the Civil War in England in 1643, the puritanical government raised duty (taxes) on both the importation of spirits from the Netherlands and alcoholic beverages that were produced at home. Scotland was not controlled by the English government at that time, and so was unaffected by these taxes. However, the growth of whisky production prompted the Scottish Parliament to pass an Act in 1644 imposing an excise duty on spirits. Collecting the taxes was very difficult, as there were few tax collectors and many of the distilleries were situated in remote and inaccessible places. In 1707, when the Act of Union brought Scotland under English legislation, further attempts were made to control whisky distillation. Many laws were passed and the situation became very confused and inconsistent with distilleries being taxed at various different rates. The tax collector, or exciseman, would be accompanied by English soldiers, called "redcoats," after the color of their uniforms. The soldiers offered some form of protection since collecting taxes was a dangerous task and several excisemen are known to have lost their lives during this period. Outwitting the redcoats became a national sport in Scotland and tales of heroism and cunning against them are woven into the history of individual distilleries. For example, distilleries were quite small and an illegal still could be broken down quite easily and removed, so detection was difficult. An illegal still comprised a metal pot in which the barley, yeast, and water were boiled over a fire, a small length of pipe cooled by water which collected the steam, and a barrel for the raw spirit.

In 1823 an Act of Parliament was passed, which made distilling legal provided a license fee was paid and the distillery produced more than 40 U.S. gallons a year. Then, in 1840 duty was imposed on each bottle sold by the distillery. Duty is still collected by the government from the distilleries on each bottle sold in the United Kingdom.

While distilleries could be hidden, it was not so easy to hide stocks of whisky. There are many tales of how precious stocks were saved from destruction or confiscation. In 1798, Magnus Eunson was distilling whisky at Highland Park in Orkney. A notorious smuggler and local preacher, he used to hide barrels of whisky in his church. On hearing that excisemen were in the area, the barrels of whisky were taken from the church and hidden under a white cloth in his house. While the excisemen searched the church, Eunson and his staff placed a coffin lid under the cloth and started a funeral service. One of Eunson's employees whispered that smallpox had been the cause of death and this was enough to make the excisemen flee in terror.

At this time, whisky was not the preferred drink for the gentry in the cities of Scotland or, indeed, England; cognac and fine wines from France were more likely served at their tables.

An illegal still at Royal Lochnagar.

Whisky pagoda and wrought iron gates—Highland Park Orkney.

The situation would change with the introduction of a license fee aimed at sorting out duties and regulations concerning the distillation of whisky in Scotland and Ireland. The first change occurred in 1823. Legalizing whisky production meant that permanent distilleries were built, leading to an improvement in quality of the finished product. The first distillery to take out a license was The Glenlivet in 1824, closely followed by Cardhu, The Glendronach, Old Fettercairn, and The Macallan among others. The earliest recorded commercial distilleries date back to the late eighteenth century, including Bowmore (1779), Highland Park (1795), Lagavulin (1784), Littlemill (1772), and Tobermory (1795).

In 1863 an outbreak of phylloxera in France started to decimate the vineyards. By 1879 most European vineyards had been forced to destroy their vines. The production of wine and, more importantly, cognac came to a halt. Customers had to look elsewhere for their liquor and attention turned to the homegrown product. At this time Adrian Usher had been experimenting with whisky-blending in Edinburgh and came up with a lighter, more palatable drink more likely to appeal to the popular taste. Many new distilleries were built during this time such as Benriach in 1898, The Balvenie in 1892, and Dufftown in 1896.

The growth in whisky sales, however, halted in 1898 when Pattisons, a well-known blending company, went bankrupt. The Pattison brothers were given prison sentences and the collapse of their company had serious repercussions throughout the industry. Under-capitalization, overspending, and a general decline in the economic climate led to the closure of many distilleries.

Finding a market for malt whisky became harder, as competition increased. It was not until 1963 that there was specific interest in single malts, however, as most were destined to satisfy the demand for blended whiskies. Several companies, notably William Grant & Sons who placed considerable resources behind The Glenfiddich, started to market single malts aggressively.

Economies of scale, high employment and marketing costs meant that many distilleries were unable to survive on their own. The growth of groups, such as United Distillers, ensured that many of these distilleries could continue producing whisky. Revival of interest in malt whisky in recent years has persuaded a few entrepreneurs to set up their own business, and a few distilleries are now in independent hands again.

The Balvenie distillery circa 1880.

A Short Whisky Thesaurus

The word *whisky* reputedly derives from the Gaelic *uisge beatha* (water of life). Whisky distillation in Scotland is now mainly in the hands of large organizations and has grown from its farming origins, although many distilleries are still small and located in rural areas. As in the past, distilleries still play an important part in community life and are often the main source of employment in country villages.

To qualify for the designation *whisky*, the spirit must be produced from water and cereals, distilled at an alcoholic strength by volume of less than 94.8%, matured in casks not greater than 185 U.S. gallons capacity, for a minimum of three years from the date of distillation, in a bonded warehouse.

For a whisky to qualify for the name *Scotch*, it has to be produced in a distillery in Scotland and matured in Scotland.

A single malt whisky is distilled at an individual distillery and produced only from malted barley. When bottled, a single malt may include whisky from several years' production from the same distillery. The age shown on the bottle reflects the length of time the youngest whisky included in the bottling has matured in the cask.

A vatted malt is produced by marrying together various malt whiskies from several distilleries. Vatted malts often reflect the distilleries from a particular region, such as Pride of the Lowlands and are labeled "Pure Malt" or "Scotch Malt Whisky." They cannot be described as a single malt.

Grain whisky is produced using a continuous distillation process. Malted and unmalted cereals, which are cooked under steam pressure, are used, and the resulting spirit is of a higher strength and matures more quickly than a malt whisky, as it has less constituent ingredients.

Single grain whiskies are the product of one grain distillery. These whiskies are sold by several companies including Whyte & Mackay (Invergordon) and United Distillers (Cameron Brig). The age shown on the bottle reflects the length of time the youngest whisky included in the bottling has matured in the cask.

Blended whisky accounts for 95 percent of Scotch whisky sales. A blended whisky is created from both single malts and grain whisky.

Blended whiskies are the perfect introduction to drinking whisky and can either be sampled on their own, on the rocks, with water, lemonade, or ginger ale; or they can make up a cocktail such as Bobbie Burns (with Benedictine) or a Rusty Nail (with Drambuie). Whisky topped off with hot water and lemon juice, sweetened with honey or sugar, with the optional addition of cloves, becomes what is known as a hot toddy to ward off a cold.

The deluxe blends contain a higher percentage of malt whiskies and often show an age on the label. As with single malts, this age reflects the youngest whisky in the blend. Deluxe blends include Johnny Walker Black Label 12 years, J & B Reserve 15 years, Dimple 15 years, and The Famous Grouse Gold Reserve 12 years.

ALCOHOLIC STRENGTH

Alcohol is now measured in the United Kingdom and Europe by percentage volume at 20°C (–5°F). In America the proof system is still operational. For example, 100° proof is equivalent to 50% alcoholic volume and 80° proof to 40%. The proof system was originally tested by putting a lit match to a mixture of the spirits with gunpowder. If it ignited then the whisky was of sufficient strength and "proved." The gunpowder would not flash if the spirit was too weak. In 1740 a hydro-meter was invented by one Mr. Clark and this was used to measure the strength of whisky. An improved hydro-meter was developed by Bartholomew Sikes and introduced in 1818. The Sikes hydrometer continued to be used until 1980, when the U.K. adopted the European method of measuring alcoholic strength by percentage volume.

Cask strength whiskies are sold at 68.5% alcohol by volume (some 120° proof).

Bottled whiskies are usually 40% alcohol by volume for home sales and 43% for export.

Distilling Malt Whisky

Malt whisky is the marriage of water, malted barley, and yeast. This apparently simple recipe belies the complexity of a drink made up of different colors, aromas, and tastes, which are produced by distilleries in various parts of Scotland.

A pure, clear, water source is the starting point for making a good single malt whisky. And as the water tumbles down from the Scottish hills or across peat bogs to the distillery, it will carry with it a little of its birthplace and travels—peat, heather,

The typical pagoda–style buildings of Glenturret distillery.

Cool, clear water on Islay for Bowmore distillery.

and granite. Whisky also needs heat from peat fires, the consummate skill of the distillery team, the magic of copper stills, maturation in oak casks, and good ventilation before it is ready for the market.

Ask any distillery manager what makes his malt different from his neighbor's and he will give you any number of reasons—the water, the type of barley, how long the barley is steeped in water, how long the barley is dried, whether peat is used in the drying process, the type of peat used, the fermentation time, the shape and size of the stills, the speed at which the raw spirit is collected, the size of cask used, the type of cask used, and the ambience in the warehouse. Other, more fanciful reasons are often given such as the normally damp, atmosphere in Scotland, the way the wind blows, or simply the magic of still and cask. It is probably fair to say that no one really knows the answer. The description of the production process given below explains the terminology associated with it and highlights some of these imponderables.

MALTING BARLEY

The production process starts with barley. All distilleries have their own sources of barley and managers liaise closely with farmers and agronomists to insure that the raw material meets their requirements.

Barley is steeped in water for a couple of days and then allowed to germinate. In a traditional distillery the wet barley is spread out by hand on a concrete malting floor for about seven days. During this period the barley is turned regularly to insure the temperature is maintained at the required level and to control the rate of germination. At some distilleries the traditional wooden shovel or shiel is still used to turn the barley. Only a handful of distilleries still malt some of their own barley; these include Bowmore and Laphroaig on Islay, Springbank of Campbeltown, and Highland Park on Orkney. The majority purchase their malted barley direct from maltings where the grain is turned mechanically in large rectangular boxes or in large cylindrical drums. The distillery manager then determines the exact length of time for each stage of the malting process.

Traditional floor maltings—turning barley by hand.

Once the required level of germination has been achieved, the natural enzymes in the barley are released. These produce soluble starch which converts into sugar during the mashing process. Germination is arrested by drying the barley either over a peat fire or with warm air. At Bowmore distillery, for example, the first 15–18 hours of the drying period occur over a peat fire. Peat is cut from the company's own peat bogs and left in the wind to dry. Barley dried over a peat fire absorbs phenols in

Cutting peat on Islay.

the peat, which give Bowmore its characteristic peaty aroma and flavor. The barley is dried for a further 48–55 hours with warm air.

Distilleries throughout Scotland use some barley dried over peat fires. However, many Speyside and Lowland distilleries produce single malt whiskies which contain no peaty influences. The reason for the high peat content in some Islay whiskies can be explained by the fact that this was traditionally the only local source of local fuel, unlike in Campbeltown and Speyside, where coal was readily available. Peat fires in the different regions of Scotland produce a range of phenols; in Islay, for example, peat is composed principally of decaying heathers, mosses, and grasses, while on the mainland old forests have decomposed to make peat.

Traditional peat-fired kiln.

MASHING

Adding water to the mash tun.

After a rest period the dried malt is ground to a fine grist. This process is known as mashing. The grist, which is part flour and part solids, is placed in a container known as a mash tun and boiling water is added. Distillery managers jealously guard their water supplies, which help to give each malt whisky its particular flavor and aroma. The shape and size of mash tuns vary but they are normally of copper and usually have a lid. The boiling water dissolves the flour and releases the sugars in the barley. The resultant liquid, now known as wort, is drawn off from the base of the mash tun through the finely slotted bottom, cooled, and passed into fermentation vessels, or washbacks. The solids, or draff (as they are known in Scotland) are removed from the mash tun and used as cattle feed.

Preparing wort in the mash tun.

FERMENTATION

Yeast is added either in liquid or solid form to the liquid wort, which has been cooled to around 70°C (158°F), in the large wooden washbacks. The yeast starts to ferment immediately and the mixture begins to give off carbon dioxide and foam. Scottish malt distillers use covered washbacks with lids that incorporate a rotating blade, which stops the foam pouring over the sides. The sugars are converted into alcohol and after some 48 hours a warm sweet peaty beer with an alcoholic content of around 7.5% is achieved.

Washbacks at Highland Park.

DISTILLATION

The spirit safe.

The fermented wort, or wash, is then piped to the still room. Traditionally, stills in a Scottish malt whisky distillery are made of copper. All spirits stills are handmade and every distillery has stills of a different shape and size. The size and shape of the still and the skill of the stillman contribute to the quality of the final spirit. Only a small part of each distillate will go on to make malt whisky. Normally malt whisky is produced after two distillations, but in some Lowland and Irish distilleries the spirit is distilled three times.

The first and largest still is the wash still, where the wash is boiled so that it breaks down into its constituent parts and the alcohol can be drawn off. The boiling point of alcohol is lower than that of water, so it is the first vapor to rise up the neck of the still. The vapor produced by the still is passed into a condenser—a series of pipe coils running through cold water. The angle of the pipe, or lyne arm, connecting the still to the condenser will affect the quality and speed of condensation.

From the condenser the liquid, now called low wines, is collected in the spirit still for redistilling via the spirit safe. It is here that the involvement of the British Customs and Excise begins. All spirits are subject to customs duty in the United

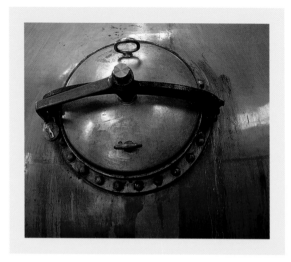

Detail of a pot still.

Kingdom and production is strictly controlled to insure that the correct amount is paid to Customs and Excise. Locks are fitted to the spirit safe by a Customs and Excise representative and measurements are taken of the spirit produced. The spirit safe contains several glass bowls into which the spirit can be directed using external faucets by the stillman. The low wines contain approximately 30% alcohol and must be distilled again in a spirit still, as they are unpalatable. Spirit stills are usually smaller than wash stills. The second distillation, which produces the pure spirit, is a carefully orchestrated and very precise procedure.

The stillman will start to test the spirit as soon as the rising vapors are condensed and pass through the spirit safe. He will direct the first liquid, which is known as foreshots, to the glass bowls, which drain off into a collecting tank at the back of the stillroom. Foreshots turn cloudy when they come into contact with water as they are still impure. The stillman can test the spirit in the spirit safe by adding water at regular intervals and checking the specific gravity.

As soon as the spirit starts to run clear, the stillman will immediately turn the faucets on the outside of the spirit safe and direct the spirit into what is known as the spirits receiver. To ensure the clarity and purity of the spirit the speed of distillation is normally reduced at this stage. The stillman will continue to check the specific gravity and clarity of the spirit. After several hours the spirit will start to weaken. This weakened spirit, or feints, is also discarded, and the stillman diverts it to the low wines and feints-collecting tank.

In some distilleries, notably Bushmills in Northern Ireland and Auchentoshan in the Lowland region of Scotland, the spirit passes through a third still to produce a lighter whisky. This is known as triple distillation.

A watery residue remains in the still once the feints have been collected. This residue is called spent lees and it is normally discharged into the sewer after treatment. At Royal Lochnagar Distillery the spent lees are sprayed over the surrounding farmland.

The foreshots and feints will be added to the next wash for distillation, when the process will be repeated.

Spirit and wash stills at Bowmore.

MATURATION

Making barrels.

The spirit is colorless, crude, and fiery—at this stage it has some of the characteristics of whisky but certainly none of its final elegance. It now has to mature in barrels for three years before it can legally be called whisky. During this time the spirit will become softer and will start to turn color as the residues of bourbon, sherry, or port in the wood casks it is stored in are absorbed.

The immature spirit is piped to the filling room, where it is poured into oak barrels. Distillery managers exercise stringent controls to ensure that the spirit is carefully measured as it is piped into barrels for maturation. All whisky warehouses are bonded; that is, they hold goods in bond by the government and every barrel is accounted for, so that the appropriate amount of duty is paid at the time of bottling.

Each distillery uses different types of barrels; at Laphroaig, for example, only barrels which once stored American bourbon are used. Other whiskies, such as The Macallan, use sherry casks, and at Glenmorangie some malt whiskies are finished in old port and old madeira casks. The type of cask used will help to determine the final color and flavor of the malt whisky. Once the barrels are filled they are stored in bonded warehouses for a minimum of three years. If they are to be used for a single malt or for a deluxe blend the barrels will be stored for at least 10 to 15 years.

As wood is permeable, the air surrounding the barrel will seep into the whisky, so if the air is salty, seaweedy, heather-scented, pine-laden, or oak-imbued, it will add to the characteristics of the malt. Some of the spirit will seep out of the barrels; most distillery managers quaintly describe this as "the angels' share." The temperature and humidity of the warehouses will also affect maturation. The longer a malt whisky is left to mature in the barrel, the more changes will take place, which is why malts of varying ages from the same distillery are so different.

From time to time the barrels are tapped to check that all is well. A firm resonant sound means that the barrel is intact and the whisky is maturing well. A leaking or broken barrel produces a dull sound and the distillery manager knows that the barrel must be inspected and probably replaced. The distillery manager will draw a small amount from the barrel and pour it into a nosing glass. He will nose (smell) the whisky and swirl it around in the glass—a "string of pearls" around the surface of the liquid shows him that the whisky is maturing satisfactorily— before returning the spirit to the barrel.

Maturing slowly in the dark; whisky for the future.

In the past, distillery workers were given old casks in which whisky had matured for many years. The casks were filled with hot water and steam and then rolled down the street. This produced several gallons of spirit. Unfortunately for today's distillery workers this practice is no longer legal.

The Malt Whisky Regions
of Scotland

J ust as the wines of France are grouped according to their region of origin, so too are the malt whiskies of Scotland. The malts in this book are listed in alphabetical order with a regional classification. Defining malt whiskies too rigidly, however, suggests uniformity; it would be inaccurate to say that all Islay malts have a strong peaty taste—Bunnahabhain does not, for example. However, there are some regional characteristics which influence the choice of a malt whisky.

Rolling hills and clear-running water are images commonly associated with whisky.

Much of the Lowland landscape is gentle with lush vegetation.

LOWLANDS

The Lowlands of Scotland with their undulating countryside do not immediately spring to mind as the home of Scotch whisky which most commonly conjures up images of mountains and tumbling streams. In this part of Scotland there are no granite hills and very few peat bogs. There is, however, a ready supply of fine barley and pure water. Lowland malts have a sweeter, mellower flavor than those from other regions, which owes much to the inherent qualities of the malted barley. Most Lowland malts are produced with very little peat. The notable exception is Glenkinchie, which is a slightly dry, smoky malt.

In the late nineteenth century there were far more malt whisky distilleries than there are today in this region. Historically the Lowland region, which is situated below an imaginary line linking the Clyde and Tay rivers, produced whisky in large industrial stills, which had none of the delicacy or range of flavors of the Highland malts. However, these are all long gone and the few remaining distilleries produce fine malts which are characteristically lighter, without the taste of peat or the sea.

Auchentoshan, which greets visitors to Glasgow arriving along the north bank of the Clyde, is the only distillery carrying out triple distillation.

Within the Lowland region are the two major cities of Scotland; Edinburgh and Glasgow, and the great shipping highway of the River Clyde. The River Clyde gave distilleries easy access to overseas markets, and ships loaded with illegal and legal stocks of whisky were a common sight. The shipyards of the Clyde built such famous ships as the *Queen Mary* and *Queen Elizabeth I* and *II*. South of the industrial area of the Clyde the landscape changes to agricultural countryside with fields of cereals and sheep grazing on the low hillsides. Here is the home of the famous Scottish cashmere and visitors to this region should make a stop at one of the many knitwear mills.

The Lowland region has much in common with Northern Ireland, the home of Bushmills. The proximity of Northern Ireland to this Scottish region led to an interchange of skills and it is believed that the Auchentoshan distillery was founded with assistance from Irish monks. One principal characteristic of Irish whiskey and the malt produced at Auchentoshan is that it is triple distilled. This extra distillation removes more of the compounds in the spirit and thus the final product is exceptionally pure.

Lowland malts are not affected by the strong winds off the sea, like island malts, and there is little salt in the finished product. The light warm breezes across the rolling countryside probably add to the final mellowness of the malts.

HIGHLANDS

As the traveler takes the road north from the Speyside region there are fewer whisky distilleries. The road will take you past the site of the now dismantled Glen Albyn distillery and to the distilleries of Glen Ord, Teaninch, Dalmore, Glenmorangie, Balblair, Clynelish, and then at the end of the road, near Wick, is Pulteney, the northernmost distillery on the mainland of Scotland. These distilleries are referred to as being situated in the Northern Highlands region. This is a mountainous part of Scotland where streams tumbling over granite, heather hills, and green glens

Rannoch Moor, Western Highlands ▶

Ben Nevis, possibly the most famous of Scotland's landmarks.

introduce interesting flavors and aromas to the malt whiskies. This large area stretches on the mainland from Pulteney in the northeast to Oban in the west, to Tullibardine in the south. Each whisky is different from its neighbor and their characteristics owe much to the local topography and water supply. As with other regions its isolation meant that few visitors made the journey to the farmhouses down narrow lanes, so illegal distilling was relatively easy. Most of the distilleries, which produce whisky today, however, were constructed in the early 1800s. Only Balblair lays claim to being built before distilling was legalized in 1790.

This is a spectacularly beautiful part of Scotland and the visitor is rewarded with exceptional views. The malt whiskies produced on the islands of Mull, Jura, and Orkney are also included in this region. There is now only one distillery left on Mull. Whisky produced here uses water that has flowed over peat bogs, which imparts a smoky flavor. On Jura the water flows over rocks and the whisky has a fresh, flowery aroma which reflects this beautiful, wild, lonely, island.

THE ORKNEY ISLES

At the northernmost tip of Scotland lie the Orkney Isles, a group of islands with the Atlantic Ocean to the west and the North Sea to the east. The Orkney Isles are nearer to Oslo than to London and it is no surprise to find many Norse influences here.

Standing stones and burial chambers from the Bronze Age can be seen at Skara Brae and Maes Howe. More recent history can be seen in the waters of Scapa Flow where wrecks of the German naval fleet in World War I still stick out of the water. Across the Churchill Barriers, which were built to protect Scapa Flow in World War II is the Italian chapel on Lamb Holm. This chapel has beautiful murals inside painted by Italian prisoners of war and is well worth a visit.

The Orkney islands are fertile with rolling countryside. Standing on the Main Island, the largest island in the group, one has a sharp sense of sea and sky, for the land is principally flat and treeless. It is difficult to say where the sky ends and the sea begins. On Hoy the land is more rugged and rises to some 1,400 feet above sea level.

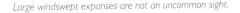

Large windswept expanses are not an uncommon sight.

The coastal cliffs are a bird watcher's paradise with curlews, kittiwakes, guillemots, puffins, and sometimes the great skua. The meadows are covered in wild flowers and the moorland and low hills are a blaze of purple when the heather comes into bloom.

The isolated location of these islands meant that whisky distilling could carry on virtually undiscovered and records show that several distilleries were built here. In 1805 excisemen destroyed a large number of distilleries on the outlying islands. Now only two remain, Scapa, which is currently mothballed, and Highland Park. Both of these distilleries are situated on the main island.

The island's natural resources—a plentiful supply of water, fertile soil for growing barley, and large supplies of peat—meant that whisky distillation could carry on undisturbed.

Orkney whiskies smell of sea air, which seeps into the wooden barrels as they mature in warehouses close to the seashore. Peat on the island, which is used to dry the malted barley, is made from heather and this imparts a honey flavor to the whisky.

A beautifully rugged landscape from Speyside.

Heather has been long associated with the malt whisky tradition.

SPEYSIDE

The Speyside whisky region sits in the Highlands of Scotland. The River Spey flows between the Ladder and Cromdale Hills into the Grampian Hills. Numerous streams flow down to the river Spey, notably the River Avon and the River Livet. There are now only two distilleries situated along the valley of the River Livet; The Glenlivet and Tamnavulin. The valley floor of the River Livet is quite wide, but it narrows steeply as the valley rises up into the hills, where narrow pathways recall the old smugglers' roads to the key cities in the Lowlands. This mountainous region was virtually inaccessible during the seventeenth and early eighteenth centuries so illegal distilling was a favorite pastime.

With its abundant supplies of fresh water, easy access to barley, and peat on the moorland hills, Speyside was a region with everything readily available for distilling whisky. Most farmers produced some whisky for their own consumption which was accepted. Trouble came when farmers started to sell whisky and the government decided that such stocks should be taxed. Most farmers refused to pay. However, one landowner, the Duke of Gordon, worked with others to introduce a law legalizing the distillation of whisky. One of his own tenants, George Smith, applied for the first license to distill whisky in 1824. Smith was a colorful character who had changed his name from Gow, as his family had long been supporters of Bonnie Prince Charlie, who was unsuccessful in his claims to the Scottish throne. Many distilleries incorporated the word "Glenlivet" in their whisky to let the consumer know that the

Vast lochs of still, clear water provide distillers with the perfect source.

whisky came from this particular part of Speyside. This led to problems of recognition and in 1880 the Smiths took others to court to stop them from using the words The Glenlivet. The judgment was found in the Smiths favor and now distilleries can only use Glenlivet by adding the word to their own name—for example, Tomintoul Glenlivet. Only one distillery can call itself The Glenlivet.

Many distilleries in this region use underground springs for their water supply. The purity of the water, which has come from deep in the hillsides, must contribute to the final product. The water flows far longer over granite hills in this part of Scotland, imparting a crispness and a different flavor to the whisky. Being away from the sea and the influence of salt laden winds, there is really no salt in the final product. The finished taste is cleaner, perhaps simpler than that of a complex Islay single malt such as Laphroaig. There is less peat here, so the traditional fuel for the fires to dry the malt was more likely to have been coal. Barley is malted with very little peat and the final spirit has less of a smoky flavor and aroma. There is more space here and maturing whiskies take on the spirit of heather and the country.

The individual distilleries throughout Speyside produce different malts, but most of them have a characteristic balance and sweetness, for Speyside malts are made using very little peat. The long history of distilling in this region has led to distilleries producing exceptionally fine malts.

This area has the highest density of malt whisky distilleries per square mile, about 40 in all, and could arguably be designated as part of the Highland area. In spite of the number of distilleries which incorporate the word Speyside into their names, very few of them are actually built on the River Spey.

CAMPBELTOWN

A century ago a visitor to Campbeltown would probably have arrived by boat. As the town and surrounding area came into view nearly 30 distillery chimneys would have been silhouetted against the skyline. For this was once the heart of whisky-making, but now, sadly, the number of distilleries is reduced to two—Springbank and Glen Scotia. The sea air of the Mull of Kintyre, on which Campbeltown is situated, imparts a special flavor to these malts.

ISLAY

The island of Islay is situated on the west coast of Scotland off the Mull Of Kintyre. Islay is the most fertile of the Hebridean Islands which run along the west coast of Scotland. The island is diamond-shaped with a deep inlet—Loch Indaal on the south west coast.

The island is steeped in history. Standing stones from the Bronze Age, old Celtic crosses dating from 800AD at Kildalton, an old medieval chapel at Kilnave, and numerous tales of illegal whisky distilling all add detail to Islay's intriguing past. The journey by boat from Kennacraig on the mainland takes some two hours. Before air travel the citizens of Islay were often cut off from the rest of the world. It is not, therefore, surprising that whisky distilling started on Islay way back in the 16th century. It is believed that the first distillers came from Ireland across the sea in the early 1500s.

Islay is a beautiful, though often windswept, island with rugged hills—the highest peak is Beinn Bheigeir, which rises to 1,450 feet above sea level—deep wooded valleys, miles of open moorland, and rolling agricultural countryside. With its isolated position, ready source of peat fuel, unlimited water, and local barley Islay was a natural place to start whisky distillation, and the seven distilleries on the island produce very different malt whiskies, from the light Bunnahabhain and Caol Ila to the stronger, aromatic Laphroaig and Ardbeg.

All distilleries on Islay are built close to the sea so that whisky could be transported easily to the mainland. Cool, clear water bubbles out of the ground, over rocks, into lonely hillside pools and then down to the sea. Barley grows on the fertile fields and peat is dug from the seemingly unlimited supply on the moorland. Peat on Islay is different from peat on the mainland of Scotland as it is composed of decaying heathers, mosses, and grasses. Malted barley dried over a peat fire in Islay has a taste and an aroma all of its own. Islay malts are known for their peatiness and several distilleries produce whiskies with a pungent peaty smell. Peat bogs cover much of Islay and the main road from Port Ellen to the capital Bowmore literally floats on the peat in certain spots. It is therefore unwise to drive too fast along this road as there are unforeseen bumps on the surface.

◄ *The familiar sight of rugged rocks and tumbling streams.*

ORKNEY ISLANDS

LEWIS

SKYE

HIGHLANDS

Inverness
Loch Ness

SPEYSIDE

Elgin

Grantown

Aberdeen

Fort William

Ben Nevis

Pitlochry

MULL

Oban

JURA

Edinburgh

Glasgow

ISLAY

Ayr

LOWLANDS

Campbeltown

ENGLAND

N
W   E
S

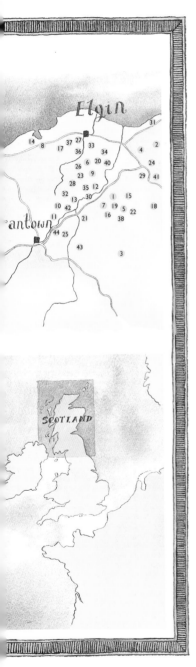

THE SINGLE MALT WHISKY REGIONS OF SCOTLAND

The map on the left shows how the distilleries in the Directory section fall into their regional groups. There are so many of them in Speyside that the area has been enlarged to make it easier to read at-a-glance.

The colors used for each region are repeated in the Directory section for ease of reference, and the distilleries have been numbered in ascending order alphabetically, as they appear in the Directory.

LOWLAND
1 Auchentoshan
2 Bladnoch
3 Glenkinchie
4 Rosebank

HIGHLAND
1 Aberfeldy
2 Ben Nevis
3 Blair Athol
4 Clynelish
5 The Dalmore
6 Dalwhinnie
7 Deanston
8 Drumguish
9 Edradour
10 Glencadam
11 Glen Deveron
12 Glen Garioch
13 Glengoyne
14 Glenmorangie
15 Glen Ord
16 Glenturret
17 Inchmurrin
18 Oban
19 Old Fettercairn
20 Royal Brackla
21 Royal Lochnagar
22 Teaninch
23 Tomatin
24 Tullibardine

MULL
25 Tobermory

JURA
26 Isle of Jura

ORKNEY
27 Highland Park
28 Scapa

SKYE
29 Talisker

SPEYSIDE
1 Aberlour
2 An Cnoc
3 Ardmore
4 Aultmore
5 The Balvenie
6 Benriach
7 Benrinnes
8 Benromach
9 Caperdronich
10 Cardhu
11 Cragganmore
12 Craigellachie
13 Dailuaine
14 Dallas Dhu
15 Dufftown
16 Glenallachie
17 Glenburgie
18 The Glendronach
19 Glendullan
20 Glen Elgin
21 Glenfarclas
22 Glenfiddich
23 Glen Grant
24 Glen Keith
25 The Glenlivet
26 Glenlossie
27 Glen Moray

28 Glenrothes
29 Glentauchers
30 Imperial
31 Inchgower
32 Knockando
33 Linkwood
34 Longmorn
35 The Macallan
36 Mannochmore
37 Miltonduff
38 Mortlach
39 The Singleton
40 Speyburn
41 Srathisla
42 Tamdhu
43 Tomintoul
44 The Tormore

ISLAY
1 Ardbeg
2 Bowmore
3 Bruichladdich
4 Bunnahabhain
5 Caol Ila
6 Lagavulin
7 Laphroaig

CAMPBELTOWN
1 Springbank

ARRAN
1 Arran

Malt Whiskies from Around the World

While Scotland may be synonymous with whisky to some, it is not the only place where whisky is made. The ingredients needed for malt whisky production are a good supply of fresh clear water, barley, peat, and a cool temperate climate, enabling the whisky to mature slowly in wooden barrels. Single malt whiskies are produced from malted barley in Ireland, both north and south, Japan, and now Tasmania.

NORTHERN IRELAND

The close proximity of Northern Ireland to Scotland and the Hebridean islands in particular has encouraged the belief that whisky distilling was brought to Scotland from Northern Ireland in the early 1500s. Recent research on the island of Rhum, situated south of the island of Skye, has indicated that this may not be true and that distilling could have started in Scotland as well some 4,000 years ago. Whatever the case may be, the influences of Northern Ireland on Scotch whisky and the influence of Scotland on Irish whiskey cannot be disputed.

There are many links between Northern Ireland and Scotland. They share a common language—Gaelic (which has since developed differently in the two countries), and the landscape is similar with its large lakes, tumbling streams, peat bogs, rolling agricultural countryside, and mountains. Indeed, legend would have it that the Giant's Causeway off the north coast of County Antrim was a footpath to Scotland. This is, of course, untrue as the Giant's Causeway was the result of erosion by the sea. Bushmills is the oldest recorded distillery and is situated at Coleraine in County Antrim on a line to the south of the Mull of Kintyre and thus further south than most distilleries in Scotland.

EIRE

The art of distilling whisky was well known in Ireland by the fourteenth century. Much of the current production is distilled using a mixture of grains and unmalted barley, oats, wheat, and rye. Several distilleries produce a single malt occasionally and these include Middleton and Cooley. Eire is south of Scotland and the more temperate climate produces whiskies with a slightly spicier taste and a crisp finish.

JAPAN

Japanese whisky owes its origins to Scotland. The first distillers were trained in Scotland and took their new-found knowledge to traditional sake distilleries in Japan. The landscape of the northern island, Hokkaido, is very similar to that of the highlands of Scotland with peat bogs, mountains, and cool, fresh streams, which flow over granite rocks. The peat produces a less intense aroma than Scottish peat. The largest company, Suntory, has distilleries at Yamazaki, near Kyoto, Hakushu, and Noheji, on the main island of Honshu. Suntory produces most of its whisky for home consumption with about three percent exported principally to the Pacific. The second biggest company is Nikka, which produces two single malt whiskies, Nikka and Yoichi. The third is Sanraku Ocean which also has two distilleries, only one of which produces malt whisky.

TASMANIA

Andrew Morrison, a farmer in Tasmania, has started to produce a single malt whisky at Cradle Mountain distillery. Tasmania has an ideal climate for producing whisky and the right local ingredients.

Enjoying Single Malt Whisky

There is a great deal of mystique surrounding the drinking of a fine single malt whisky. It is often regarded as a male prerogative. Pictures of old men hiding behind newspapers in their clubs, or of connoisseurs expounding the virtues of one particular malt over another are perpetuated in cartoons and articles about whisky.

There is also some debate as to when you should drink whisky. Generally speaking, a whisky can be drunk at any time, although we may not wish to emulate the people of the Hebrides, of whom John Stanhope, a Scottish businessman who traveled throughout Scotland, wrote in his diary of 1806: "…they still continue to take their streah, or glass of whisky, before breakfast, which, though by no means a palatable regimen to Englishmen, seems at least to be a very wholesome one, if one may judge from the healthy appearance and ruddy skins of the natives—indeed, in such a wet climate, it is almost absolutely necessary to drink spirits in some degree. Additional streahs are never refused in the daytime."

Perhaps we should follow the advice of W. C. Fields, who is reported to have said: "I gargle with whisky several times a day and haven't had a cold in years."

Throughout the Directory it is suggested that certain malts are best served before dinner and others after dinner. These are the personal guidelines of the author. Experimentation is the best way to discover one's own preferences.

STARTING A WHISKY COLLECTION

The wide range of single malt whiskies available offers the whisky drinker a voyage of discovery. For the dedicated whisky collector the choices are infinite. In addition to the malts released by the individual distilleries, there are bottlings from specialists at different ages and strengths. Starting a whisky collection is relatively easy—many wine merchants and liquor stores stock a variety covering the principal regions of Scotland. Irish malt whiskey will usually also be available.

Everyone's taste is different, but the following ten single malt whiskies provide a good introduction to the different regions:

BOWMORE, 17 years old, *from Islay*

LAPHROAIG, 10 years old, *from Islay*

HIGHLAND PARK, 12 years old, *from Orkney*

TALISKER, 10 years old, *from Skye*

GLENKINCHIE, 10 years old, *from the Lowlands*

THE BALVENIE DOUBLE WOOD, 12 years old, *from Speyside*

BENRIACH, 10 years old, *from Speyside*

THE SINGLETON OF AUCHROISK, 10 years old, *from Speyside*

EDRADOUR, 10 years old, *from the southern Highlands*

GLENMORANGIE, 12 years old sherry wood finish, *from the Highlands*

This collection offers a choice of pre-dinner and after-dinner malts. (See the Directory for further information on each.)

STORING WHISKY

Once whisky has been removed from the cask and placed in a bottle, where the air cannot get to it, the aging process ceases. As with most spirits, it will retain its color, aroma, and flavor sealed in the bottle. Once it has been opened, there is no need to drink it all immediately, as with a fine port or sherry. There may be some loss through evaporation, particularly if the cork does not fit tightly enough. If the bottle remains open for a very long time there will probably be some imperceptible changes in aroma and taste.

Store opened bottles upright with the cork firmly replaced at room temperature, and away from the light.

Tasting Single Malt Whisky

A single malt whisky is a drink to be savored. The following notes will help you to learn more about malt whiskies and to understand the tasting notes in the Directory.

Look at the color. Every malt has its own particular color, from palest gold to the darkest brown. This is the result of the maturation process, which takes place in plain oak, old bourbon, sherry, port, or madeira casks, each one imparting its color and flavor to the whisky as it slowly matures.

Pour a little into a glass, and cover the glass with your hand for a little while. Move your hand slightly away to uncover the scent of the whisky. The whisky blender uses a nosing glass to smell or "nose" a single malt. Similar in style to a sherry glass, this confines the vapors and makes it easier to identify the different aromas.

Take your hand completely away from the top of the glass then swirl the whisky around. This will release other fragrances.

Sip the malt slowly. As you roll it around your tongue, flavors will be introduced into the mouth. Distinct taste sensations will be experienced in different parts of your mouth.

Note how the taste changes as you swallow the malt whisky. This is known as the "finish." Some single malts will have a more pronounced range of aftertastes than others. When you have discovered the full impact of the aromas and tastes of your chosen malt whisky you may choose to add water to the glass. This will also help to release the flavors. Ideally, the water should come from the same source as the distillery but this is usually impractical, so bottled water, preferably still, is recommended.

ASSESS THE COLOR OF THE WHISKY.

SWIRL THE WHISKY TO RELEASE THE AROMAS.

COVER THE GLASS TO CONTAIN THE AROMAS.

NOSE THE WHISKY.

TASTE THE WHISKY.

Glassware

A wide variety of glassware is available to the whisky drinker, and a few of the more popular choices are listed below.

Crystal Traditionally, whisky is drunk from a small cut-glass tumbler. A cut-glass decanter filled with whisky surrounded by several tumblers can be very beautiful with the refracted light changing the whisky from golden to ruby to amber. Edinburgh Crystal has been making such crystal for over 125 years, but its origins can be traced back to the seventeeth century.

Plain glass Some whisky drinkers prefer to use a plain glass so that they can see the color more clearly. Professional tasters use a nosing glass which has a distinctive shape. A nosing glass is supplied with a glass cover, which allows the fragrances of the malt to be captured inside the glass.

Quaich Centuries ago the favorite drinking cup throughout Scotland was the quaich (from the Gaelic *cuach*, shallow cup). Originating in the western Highlands, certain sizes became reputed whisky measures and one of these was generally used when offering the cup of welcome to the visitor and serving the farewell or parting cup. The primitive wood form was superseded by horn and then silver. Its simple shape with its two handles or ears, colloquially known as lugs remains unchanged. (Quote from Hamilton & Inches Ltd., Edinburgh.)

Edinburgh crystal.

For the whisky drinker, a quaich and hip flask.

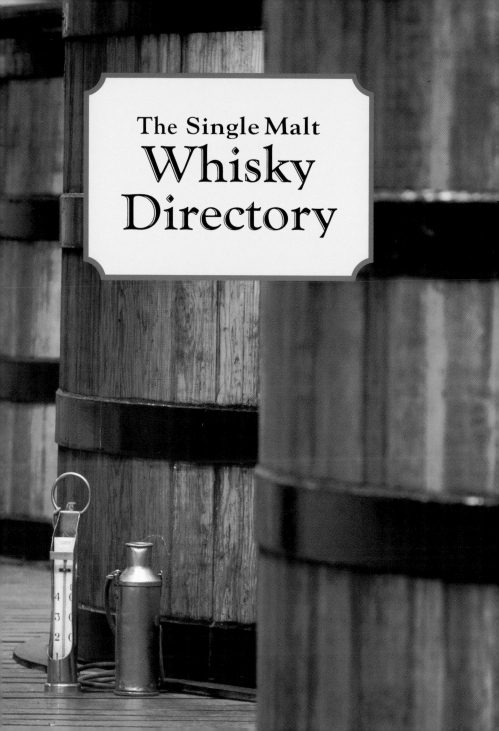

The Single Malt
Whisky
Directory

A Guide to The Directory

An old Scottish lament states that "A whisky bottle's an awful inconvenient thing: it's owr muchle for ane, an'nae eneuch for twa!" This means that a whisky bottle is too much for one to drink, but not enough for two. I would counsel readers of this book to taste the whiskies described with far more caution.

Follow the notes on tasting a single malt whisky on pages 46–47 and you will gain much pleasure from sampling the different malts available. For newcomers to single malt a drop of water should be added, as this reduces the intensity of flavor and allows the palate to savor the different tastes. Drinking a cask strength malt whisky without adding any water can be quite a shock to the system and could put you off Scotch for life. Start with the softer Lowland malts, graduate to Speyside and then try malts from Orkney, Mull, Jura, and Islay. This way you will come to learn the differences in each malt and appreciate the flavors of heather, peat, moorland, and sea.

The first section of the Directory is a listing of distilleries whose single malts are available worldwide. Most of the whisky distilleries are fully operational, but some have been closed down for a while and could reopen any time. The Scotch whisky industry talks of such distilleries as being mothballed; everything is kept in pristine condition until the day comes for production to start again.

Whiskies are included from Scotland, Northern Ireland, and Japan. Unfortunately, not all whiskies were available for tasting. In Japan, there are at least a dozen malt whisky distilleries, but most of their production is sold at home and very little is exported. In Eire, at the Cooley distillery two single malts are produced, Tyrconnell and Connemara, and in Tasmania, Castle Mountain single malt whisky is now available.

The whiskies are in alphabetical order and listings show which country they come from and, in the case of Scotland, the region in which each distillery is located. Each region has a color to tie in with the map on pages 40-41 for easy-reference.

The introduction on each page gives the key information about the distillery, historical notes, when it was founded and where it is situated. The distillery address, telephone numbers, and where possible fax numbers are also given.

DISTILLERY FACTS

At-a-glance facts are listed for each entry. They will always be in the same order, and will be given using the symbols below. Any information that was not available at the time of writing has been marked N/A.

🥃 = Founded 🏛 = Owners 🖐 = Distillery manager 〰 = Water supply

⛏ Stills 🛢 Casks 🛈 Visiting information

AGES, BOTTLINGS, AWARDS

This box gives information of the ages available. The age signifies the number of years the whisky has matured in the barrels before bottling. On average a single malt whisky is matured for 10 years before it is bottled, but as the reader will observe, other ages are common.

Special Bottlings are most often available overseas particularly in Korea and Japan. It is, however, worth talking to your local specialist retailer to see whether any are available.

Awards are mentioned where applicable. The IWSC Awards relate to awards given each year by the International Wines & Spirits Competition.

TASTING NOTES

Each brand is accompanied with tasting notes. The information here covers age, nose, and taste, all based on the author's personal opinion.

Aberfeldy

ABERFELDY DISTILLERY, ABERFELDY, PERTHSHIRE PH15 2EB
TEL: +44 (0)1887 820330 FAX: +44 (0)1887 820432

ABERFELDY DISTILLERY was founded in 1896 by John Dewar & Sons Ltd. on land belonging to the Marquis of Breadalbane. The distillery was built just outside of Aberfeldy's town limits and on the south bank of the River Tay. Long associated with whisky distilling, Aberfeldy's main water source is the Pitilie Burn, a brook which also supplied another distillery until 1867. The distillery closed in 1917 along with many others, when the government decided to reserve barley stocks for food. Aberfeldy reopened in 1919, but was to close again during World War II until 1945. In 1972–73 the distillery was rebuilt and outfitted with four new steam-heated stills.

The label on a bottle of Aberfeldy single malt depicts a red squirrel; there is a colony of these animals nearby. This malt is a beautiful sun gold color streaked with red.

distillery facts	
🌿	1896
🏴	United Distillers
✍	G. Donoghue
〰	Pitilie Burn
🗲	2 wash 2 spirit
🛢	N/A
ℹ	Easter–Oct. Mon.–Fri. 10:00-4:00

ages, bottlings, awards
Aberfeldy 15 years 43%
from the distillery

tasting notes

AGE: 15 years 43%

NOSE: Warm; sherry and nutmeg.

TASTE: Medium-bodied with a hint
of smoke.

HIGHLAND
SINGLE MALT
SCOTCH WHISKY

ABERFELDY

distillery was established in
1898 on the *road* to *Perth* and
south *side* of the *RIVER TAY.*
Fresh *spring water* is taken
from the nearby *PITILIE
burn* and used to produce this
UNIQUE single MALT &
SCOTCH WHISKY with its
distinctive PEATY nose.

AGED **15** YEARS

Distilled & Bottled in *SCOTLAND.*
ABERFELDY DISTILLERY
Aberfeldy, Perthshire, Scotland.

43% vol 70 cl

Aberlour

ABERLOUR DISTILLERY, ABERLOUR, BANFFSHIRE AB38 9PJ

TEL: +44 (0)1340 871204 FAX: +44 (0)1340 871729

THE GAELIC translation of Aberlour is "Mouth of the Chattering Burn." It may have a connection with the well, found in the grounds of the distillery, and which dates back to a time when the valley was occupied by a druid community. It may have been the purity of this spring that led James Fleming to found the distillery in 1879. The distillery changed hands several times until 1945 when it was purchased by S. Campbell & Sons Ltd. It is now managed by Campbell Distillers, the U.K. subsidiary of Pernod-Ricard. Aberlour is situated at the foot of Ben Rinnes not far from the Linn of Ruthie which tumbles down 30 feet into the Lour Burn. Aberlour is a beautiful amber single malt.

distillery facts

- 1826
- Campbell Distillers Ltd.
- Alan J. Winchester
- Springs on Ben Rinnes
- 2 wash 2 spirit
- N/A
- No visitors

tasting notes

AGE: 10 years 40%

NOSE: Heady malt and caramel aroma.

TASTE: Medium-bodied with hints of peat and honey.

An Cnoc

SPEYSIDE

KNOCKDHU DISTILLERY, KNOCK, BY HUNTLY, ABERDEENSHIRE AB5 5LJ
TEL: +44 (0)1466 771223 FAX: +44 (0)1466 771359

KNOCKDHU DISTILLERY was built in 1893 for Haig's, when springs containing pure, clear, crystal water were discovered running down the southern slopes of Knockdhu—also known as the Black Hill. This, together with good local supplies of barley and peat, provided the raw materials for producing malt whisky. Production began in October 1894 and as many as 3,000 U.S. gallons were distilled each week from the two steam-driven pot stills. Most of the distillery's production was

distillery facts

- 1893
- Inver House Distillers Ltd.
- S. Harrower
- Springs at the foot of Knockdhu
- 1 wash 1 spirit
- Oak hogsheads
- No visitors

Knockdhu
SINGLE HIGHLAND MALT SCOTCH WHISKY
Established 1894

destined for blended whisky and very little of this fine single malt was available until the distillery was purchased by Inver House Distillers Ltd. in 1988 and relaunched, using the name An Cnoc.

An Cnoc is a very pale gold malt whisky and is bottled by Inver House at 12 years old.

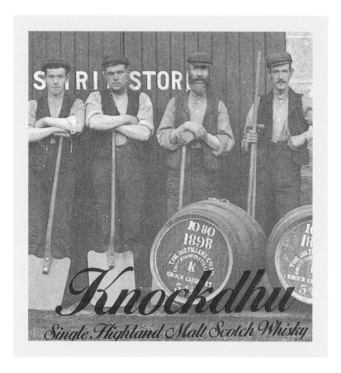

ages, bottlings, awards
An Cnoc 12 years 40%

tasting notes

AGE: 12 years 40%

NOSE: Soft, very aromatic with a hint of vanilla ice cream and smoke.

TASTE: A clean malt with a full range of fruit flavors from dry citrus to warm tropical, with a long, smooth finish. A malt for every occasion.

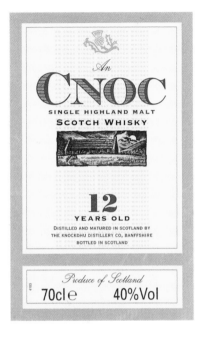

Ardbeg

ISLAY

ARDBEG DISTILLERY, PORT ELLEN, ISLE IF ISLAY PA42 7EB
TEL: +44 (0)1496 302418 FAX: +44 (0)1496 302496

AS THE Island of Islay slips into view from the sea, long low distillery buildings appear on the shore. These are known as the Kildalton distilleries and the farthest to the east is Ardbeg.

Distilling at Ardbeg started around 1798, but it was only in 1815 that the MacDougall family began commercial distilling. By 1886 Ardbeg was employing 60 people out of a total population of 200 in the village, and producing 300,000 U.S. gallons of pure alcohol a year. In 1990 Ardbeg became part of Allied Distillers Ltd.

Unfortunately, at the time of this writing Ardbeg is silent. However, supplies of this pale straw malt are readily available through specialist retailers.

distillery facts

- 1798
- Allied Distillers Ltd.
- Iain Henderson
- Lochs Arinambeast and Uigedale
- 1 wash 1 spirit
- Refill
- No visitors

tasting notes

AGE: 1974 40%

NOSE: A full, peaty aroma, slightly medicinal.

TASTE: Smoky, rich with an excellent, rounded finish. Worth seeking out.

Ardmore

ARDMORE DISTILLERY, KENNETHMONT, HUNTLY,
ABERDEENSHIRE AB54 4NH

TEL: +44 (0)1464 831213 FAX: +44 (0)1464 831428

ARDMORE DISTILLERY was built by William Teacher & Sons in 1898 at Kennethmont. This area is near the River Bogie which is at the edge of the Grampians, the range of low mountains across central Scotland separating the Lowlands and the Highlands. In 1955 the number of stills was increased from two to four, and in 1974 to eight. Many of the original distillery features are preserved on site including coal-fired stills and a steam engine.

Now owned by Allied Distillers Ltd., the bulk of the distillery's production is destined for blended whisky, in particular Teacher's Highland Cream. Special bottlings of this pale gold-colored malt with a sweet, full taste are sometimes available.

distillery facts

- 1798
- Allied Distillers Ltd.
- Iain Henderson
- Lochs Arinambeast and Uigedale
- 1 wash 1 spirit
- Refill
- No visitors

tasting notes

AGE: 1981 40%

NOSE: Sweet, with promise.

TASTE: Strong, malty, yet sweet on the palate with a dry finish. A good after-dinner malt.

Arran

ARRAN DISTILLERY, LOCHRANZA, ARRAN, ARGYLL, KA27 8HJ
TEL: +44 (0)1770 830624 FAX: +44 (0)1770 830364

IF ANYONE ever doubted the future of malt whisky distilling in Scotland, it was certainly not the Currie family who founded Isle of Arran Distillers in 1994. Arran has a history of distilling and the whisky produced always had a good reputation. The last distillery to operate here, the Lagg distillery, was closed in 1837 and it is well over a hundred years since whisky was distilled on the island.

Arran is family-owned and the Curries have built a new distillery in a traditional design in the village of Lochranza. The distillery is situated in a valley bordered by hills, close to a fourteenth-century castle and with the Eason Biorach stream running alongside, providing a source of pure water.

To create interest in Arran, the company launched a campaign whereby members of the public could purchase a bond entitling them to a case of twelve

distillery facts

- 1994
- Isle of Arran Distillers Ltd.
- Gordon Mitchell
- Eason Biorach
- 1 wash 1 spirit
- Ex-sherry hogsheads and butts
- **i** Open all year 10:00-6:00 Guided tours, audiovisual show, exhibition, shop, and restaurant

bottles of Isle of Arran Founder's Reserve single malt whisky in the year 2001. Bond holders are also members of the Isle of Arran Malt Whisky Society, which allows them to buy special blends and malts from the company.

The first spirit ran from the stills in June 1995 and is now maturing in warehouses on site in sherry hogsheads. The spirit cannot be called whisky until June 1998 and the company believes the first bottlings will be available in 2001, when the spirit will have reached some maturity. First indications are that the malt will have a peaty flavor with sweet overtones. While Isle of Arran Distillers wait for their first whisky to mature they are marketing several brands, notably this vatted malt. Eileandour is Gaelic for island water and is a blend of malts from the Highlands and Islands of Scotland.

Visitors to this very beautiful island should make a point of visiting this new distillery.

tasting notes

AGE: First Production 1995 63.5%

NOSE: Softer and sweeter than you would expect of a new spirit.

TASTE: Slightly raw with complex flavors of malt and spice.

AGE: 1 Year Old Spirit 1996 61.5%

NOSE: Less raw aroma with a little sherry and peat.

TASTE: There's a hint here of the whisky to come. It is still immature but the taste has broadened with elements of malt, pepper, honey, and peat with a sweet aftertaste.

AGE: Eileandour, 10 years old

NOSE: Sherry with peat overtones.

TASTE: Full, slightly strong on the mouth at first, then hints of vanilla and honey take over. A long mellow aftertaste.

Auchentoshan

LOWLAND

AUCHENTOSHAN DISTILLERY, DALMUIR, DUNBARTONSHIRE G81 4SG
TEL: +44 (0)1389 878561 FAX: +44 (0)1389 877368

VISITORS TO the northern edges of Glasgow can catch a glimpse of the Auchentoshan Distillery, which is situated between the Kilpatrick Hills and the River Clyde. Auchentoshan was built in 1800 and passed through many hands until 1969 when it was sold to Eadie Cairns who re-equipped the distillery. In 1984 the distillery was purchased by Morrison-Bowmore distilleries, owners of Glen Garioch and Bowmore.

Auchentoshan gives a real insight into the making of a Lowland malt. Here the whisky is still triple-distilled and fermentation is in both larch and stainless-steel washbacks. In the late nineteenth century there were many malt whisky distilleries in this region, but now there remain only six of which four are mothballed. Only Auchentoshan and Glenkinchie are fully operational at this time.

Auchentoshan has a fresh, slightly lemony aroma with a warm color, reminiscent of sunny wheatfields.

distillery facts

- 1800
- Morrison-Bowmore Distillers Ltd.
- Stuart Hodkinson
- Loch Cochno
- 1 wash 1 intermediate 1 spirit
- Ex-bourbon and sherry
- No visitors

ages, bottlings, awards

Morrison-Bowmore currently market
Auchentoshan in two versions, unaged
and at 10 years
1992 IWSC Gold Medal (21 years)
1994 IWSC Gold Medal (21 years)

tasting notes

AGE: Unaged 40%

NOSE: Warm, slightly citrusy.
Inviting.

TASTE: Very smooth fruit flavors
with a definite aftertaste.
A malt to savor at any time
before and after dinner.

AGE: 10 years 40%

NOSE: A fresh aroma with citrus
and raisins.

TASTE: Soft sweetness with a hint
of oak and lemon and a
long, round aftertaste.

Aultmore

AULTMORE DISTILLERY, KEITH, BANFFSHIRE AB55 3QY
TEL: +44 (0)1542 882762 FAX: +44 (0)1542 886467

AULTMORE DISTILLERY was founded in 1896 by Alexander Edward, who also owned the Benrinnes Distillery. The distillery sits on Auchinderran Burn; Aultmore is Gaelic for Big Burn. In 1898 Edward purchased the distillery at Oban and launched the Oban & Aultmore-Glenlivet Distilleries Ltd. Other key members of the company were Messrs. Greig & Gillespie, whisky blenders in Glasgow, and Mr. Brickmann, a whisky broker who worked with Pattisons Ltd., whisky blenders in Leith, Edinburgh. Unfortunately Pattisons went into liquidation in 1899 and forced distillation at both distilleries to be reduced. In 1923 Aultmore was purchased by John Dewar & Sons Ltd. Aultmore was one of the first distilleries to treat its waste so that it could be used as animal feed.

distillery facts

- 1895
- United Distillers
- Jim Riddell
- Auchinderran Burn
- 2 wash 2 spirit
- N/A
- No visitors

ages, bottlings, awards

Aultmore 12 years 43%

RARE MALTS
SELECTION

Each individual vintage has been specially selected from Scotland's finest single malt stocks of rare or now silent distilleries. The limited bottlings of these scarce and unique whiskies are at natural cask strength for the enjoyment of the true connoisseur.

NATURAL
CASK STRENGTH
SINGLE MALT
SCOTCH WHISKY

AGED **21** YEARS

DISTILLED 1974
AULTMORE
DISTILLERY
ESTABLISHED 1895
KEITH, BANFFSHIRE

PRODUCED AND BOTTLED
IN SCOTLAND
LIMITED EDITION
BOTTLE

tasting notes

AGE: 12 years 43%

NOSE: Delicate, summery, with a hint of honey and smoke.

TASTE: Well-rounded malt with a warm, smooth, slightly buttery taste.

The Balvenie

SPEYSIDE

WILLIAM GRANT & SONS LTD, THE BALVENIE DISTILLERY,
DUFFTOWN, KEITH, BANFFSHIRE AB55 4DH
TEL: +44 (0)1340 820373 FAX: +44 (0)1340 820805

THE BALVENIE Distillery occupies a site near the ancient Balvenie Castle and was built alongside Glenfiddich distillery by William Grant in 1892. Both distilleries still belong to the same family firm. Balvenie remains one of the most traditional distilleries in Scotland, using wherever possible barley grown nearby and its own floor maltings. The stills have remained unchanged over the last century.

The Balvenie distillery produces three distinctive malts, two of which are The Balvenie Founders Reserve, 10 years old and The Balvenie Double Wood, 12 years old. The latter is matured in traditional oak and sherry oak casks. The Balvenie Single Barrel, 15 years old, is a limited edition of some 300 bottles from a single cask. William Grant & Sons Ltd. sell The Balvenie in a special miniature gift pack, which is a great introduction to these special malts.

The Balvenie single malts vary in color from pale straw through golden honey to deep amber with a hint of copper.

distillery facts

- 1892
- William Grant & Sons Ltd.
- Bill White
- The Robbie Dubh springs
- 4 wash 4 spirit
- Oak—Spanish sherry and American bourbon
- i No visitors

tasting notes

AGE: 10 years Founders Reserve 40%

NOSE: Smoky, with citrus and a hint of honey.

TASTE: A dry, refreshing malt with a rounded taste and a touch of sweetness from the sherry casks, which lingers in the mouth.

AGE: 12 years Double Wood 40%

NOSE: Glorious, rich.

TASTE: Full-bodied, smooth on the palate with a fuller sweeter finish. A good after-dinner malt.

AGE: 15 years Single Barrel 50.4%

NOSE: Pungent and dry with a little sweetness.

TASTE: Fifteen years' maturation creates a rich mellow malt with a full caramel aftertaste.

The Balvenie Distillery

EST 1892

PROPS: WILLIAM GRANT & SONS LTD.

Ben Nevis

BEN NEVIS, LOCH BRIDGE, FORT WILLIAM PM33 6TJ
TEL: +44 (0)1397 702476 FAX: +44 (0)1397 702768

BEN NEVIS is the only distillery to obtain its water from Britain's highest mountain. It was built in 1825 by one John Macdonald, known as "Long John," whose name is still linked with whisky today. An article in the *Illustrated London News* journal in 1848 recorded Queen Victoria's visit to the distillery. Ben Nevis continued to grow and in 1894 the West Highland Railway was officially opened providing a cheap means of transporting coal to the distillery.

In 1955 the distillery was sold by the Macdonald family to Joseph Hobbs. He installed a Coffey still (see pp.242–243) and for a while Ben Nevis was one of the first distilleries in Scotland to produce both malt and grain whiskies. The Coffey still was removed some years ago. After several more changes of ownership and a silent period, Ben Nevis was purchased by the Nikka Whisky Distilling Company Ltd. of Japan in 1989, thus ensuring the future of distilling in Fort William.

distillery facts

- 1825
- The Nikka Whisky Distilling Co. Ltd.
- Colin Ross
- Alt a Mhulin on Ben Nevis
- 2 wash 2 spirit
- Mix of fresh sherry and bourbon, refill sherry and hogsheads
- Jan.–Oct., 9:00–5:00

BEN NEVIS DISTILLERY
── ESTABLISHED 1825 ──

ages, bottlings, awards
Only a small amount of Ben Nevis is
bottled each year, usually at cask
strength—ages 19, 21, and 26 years

tasting notes

AGE: 1970 26 years old
Cask No. 4533 52.5%

NOSE: Very fragrant with a sweet,
full malt aroma.

TASTE: A full-bodied malt, very
flavorful—sherry, caramel,
and peat—with a long,
sweet finish
An exceptional after-dinner
malt.

Benriach

BENRIACH WAS founded in 1898 by John Duff, who also started Longmorn Distillery, which is less than a quarter of a mile away. The two distilleries were originally linked by a railway line and the company's own steam locomotive, known as The Puggy, transported coal, barley, peat, and barrels between them. Benriach Distillery only produced whisky for a couple of years before it closed in 1900, although the maltings continued to provide malted barley for Longmorn.

distillery facts

- 🌾 1898
- 🏭 Seagram Distillers Plc.
- ✍️ Bob MacPherson
- 〰️ Local springs
- 🅰️ 2 wash 2 spirit
- 📖 N/A
- ℹ️ By appointment only

1 8 9 8

BENRIACH DISTILLERY
EST.1898
A SINGLE
PURE HIGHLAND MALT
Scotch Whisky

Benriach Distillery, in the heart of the Highlands,
still malts its own barley. The resulting whisky has
a unique and attractive delicacy

PRODUCED AND BOTTLED BY THE

BENRIACH

DISTILLERY CO
ELGIN, MORAYSHIRE, SCOTLAND, IV30 3SJ
Distilled and Bottled in Scotland

AGED 10 YEARS

70 cl ℮ 43%vol

The Longmorn Distilleries Co. Ltd. reopened the Benriach Distillery in 1965 and in 1978 was purchased by Seagram Distillers Plc. Since then Benriach has been an important part of the company's brands; 100 Pipers; Queen Anne, and Something Special. In 1994 Benriach was released for this first time as a 10-year-old single malt and is marketed by Seagrams as part of "The Heritage Selection." The distillery still continues to malt barley in the traditional manner.

Benriach single malt is a pale, honey-colored single malt.

ages, bottlings, awards
Benriach 10 years 43%

tasting notes

AGE: 10 years 43%

NOSE: Elegant, delicate aroma with a hint of summer flowers.

TASTE: Light, soft with a wide range of sweet fruit flavors, and a dry, clean aftertaste with a hint of peat A warming malt for the cocktail, before and after dinner.

Benrinnes

BENRINNES DISTILLERY, ABERLOUR, BANFFSHIRE AB38 9NN

TEL: +44 (0)1340 871215 FAX: +44 (0)1340 871840

THE CURRENT Benrinnes Distillery was founded in 1835, but records show that distilling was started here in 1826 by Peter McKenzie. The distillery was originally named Lyne of Ruthrie by John Innes who, forced into bankruptcy, sold the farm and out-buildings for distilling to William Smith. He changed the name to Benrinnes but unfortunately he, too, was forced to sell

distillery facts

- 1835
- United Distillers
- Alan Barclay
- Scurran and Rowantree Burns
- 2 wash 2 spirit
- N/A
- No visitors

and David Edward took over the distillery. The distillery was inherited by his son, Alexander, and in 1922 Benrinnes was purchased by John Dewar & Sons Ltd.

Benrinnes is built 700 ft. above sea level and obtains its water from the granite hillside. Alfred Barnard wrote in 1887 that the water "rises from springs on the summit of the mountain and can be seen on a clear day some miles distant, sparkling over the prominent rocks on its downward course, passing over mossy banks and gravel, which perfectly filters it."

ages, bottlings, awards

Benrinnes 15 years

Benrinnes 21 years distilled 1974

60.4% limited edition from United

Distillers Rare Malts Selection

1994 ROSPA Health & Safety

Gold Award

tasting notes

AGE: Benrinnes 21 years 1974
60.4%

NOSE: Rich butterscotch aroma.

TASTE: Full-bodied malt with a hint
of vanilla and fruit, slightly
oily texture, and a warm,
lingering finish.
A beautiful single malt.

RARE MALTS
SELECTION

Each individual vintage has been specially selected from Scotland's finest single malt stocks of rare or now silent distilleries. The limited bottlings of these scarce and unique whiskies are at natural cask strength for the enjoyment of the true connoisseur.

NATURAL
CASK STRENGTH
SINGLE MALT
SCOTCH WHISKY

AGED **21** YEARS

DISTILLED 1974

BENRINNES
DISTILLERY
ESTABLISHED 1826
ABERLOUR, BANFFSHIRE

60.4%vol 70cl e

PRODUCED AND BOTTLED
IN SCOTLAND
LIMITED EDITION

BOTTLE N⁰ 8503

Benromach

BENROMACH DISTILLERY, FORRES, MORAYSHIRE IV35 0EB

TEL: +44 (0)1343 545111 FAX: +44 (0)1343 540155

BENROMACH DISTILLERY was built in 1898 by Duncan McCallum of Glen Nevis Distillery and F. W. Brickman, spirit dealer of Leith, Edinburgh. The distillery has had a checkered career, for it closed almost immediately and then reopened in 1907 as Forres with Duncan McCallum still at the helm. Revived again as Benromach after World War I, the distillery was silent from 1931 to 1936, then purchased in 1938 by Associated Scottish Distillers. The distillery was rebuilt in 1966 and was then closed in 1983 by United Distillers.

In 1992 Benromach was purchased by Gordon & MacPhail. The distillery will not be fully operational until 1998.

distillery facts

- 1898
- Gordon & MacPhail
- Non-operational
- Chapelton Springs
- 1 wash 1 spirit
- N/A
- No visitors

tasting notes

AGE: 12 years 40%

NOSE: Light, sweet, and fresh.

TASTE: A good, rounded malt, light caramel with spice and a long, slightly strong finish.

Bladnoch

BLADNOCH, WIGTOWNSHIRE DG8 9AB

TEL: +44 (0)1988 402235

LOCATED AT Scotland's southern-most edge, this lowland distillery was founded in 1817 by Thomas McClelland. The distillery re-mained in family ownership until 1938 when it closed.

After several changes of owner-ship, Bladnoch was revived in 1956 and is now part of the United Distillers single malt whisky portfolio. Bladnoch was mothballed in 1993.

Bladnoch is now only available from Gordon & MacPhail. With a pale warm amber color, Bladnoch is a fine Lowland malt.

distillery facts

- 1817
- United Distillers
- Non-operational
- Loch Ma Berry
- 1 wash 1 spirit
- N/A
- No visitors

tasting notes

AGE: 1984 40%

NOSE: A sweet, delicate aroma.

TASTE: At first sweet, light on the tongue, then the malt expands into rich flavors of citrus, cinnamon, and flowers.

Blair Athol

BLAIR ATHOL DISTILLERY, PITLOCHRY, PERTHSHIRE PH16 5LY
TEL: +44 (0)1796 472161 FAX: +44 (0)1796 473292

BLAIR ATHOL Distillery was founded in 1798 by John Stewart and Robert Robertson. The distillery was revived in 1825 by John Robertson and passed through several hands until it was inherited by Elizabeth Conacher in 1860. In 1882 Blair Athol was then purchased by Peter Mackenzie who was a Liverpool wine merchant. Mackenzie was born in Glenlivet. In 1932 the distillery closed and was acquired by Arthur Bell & Sons Ltd. but it did not reopen until 1949. In 1973 the number of stills at Blair Athol was increased from two to four.

The label on a bottle of Blair Athol single malt depicts an otter. The distillery's water supply is the Allt Dour Burn or The Burn of the Otter. Blair Athol is a warm amber single malt.

distillery facts

🥃	1798
🏛	United Distillers
🖎	Gordon Donoghue
〰	Allt Dour Burn
🄰	2 wash 2 spirit
🛢	N/A
ℹ	Easter–Sept. Mon.–Sat. 9:00–5:00; Sun. 12:00–5:00; Oct–Easter Mon.–Fri. 9:00–5:00; Dec.–Feb. tours by appointment only

ages, bottlings, awards
Blair Athol 12 years 43%

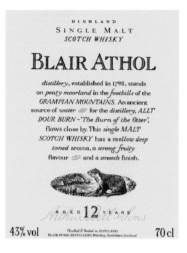

HIGHLAND
SINGLE MALT
SCOTCH WHISKY

BLAIR ATHOL

distillery, established in 1798, stands
on *peaty moorland* in the *foothills* of the
GRAMPIAN MOUNTAINS. An ancient
source of *water* 🖋 for the *distillery, ALLT
DOUR BURN* ~ '*The Burn of the Otter*',
flows close by. This *single MALT
SCOTCH WHISKY* has a *mellow deep
toned* aroma, a *strong fruity*
flavour 🖋 and a *smooth* finish.

AGED **12** YEARS

43% vol   70 cl

Distilled & Bottled in SCOTLAND.
BLAIR ATHOL DISTILLERY, Pitlochry, Perthshire, Scotland.

tasting notes

AGE: 12 years 43%

NOSE: A cold hot toddy—fresh, honey and lemon.

TASTE: A warm malt with a hint of sweetness and smoke.

Bowmore

ISLAY

BOWMORE DISTILLERY, BOWMORE, ISLAY, ARGYLL PA43 7JS
TEL: +44 (0)1496 810441 FAX: +44 (0)1496 810757

THERE CAN be no finer place to sit than on the pier at Bowmore with the distillery behind you and the blood-red sun setting out to sea. According to a Scottish legend, the sea is red at Bowmore because when the giant Ennis (Angus) was crossing Loch Indaal his dogs were killed by a dragon who had been woken from his slumbers.

Visitors approaching the distillery by road will see the Round Church which was built in 1767 by Daniel Campbell. It is a very fine building with an octagonal tower, and sits at the top of the main high street which runs through the town and to the sea. The locals of Bowmore say the church is round so that the devil has nowhere to hide.

Like all Islay distilleries, Bowmore Distillery is built close to the seashore. At Bowmore, however, a warehouse is built below sea level and the Atlantic waves break against the thick walls imparting a special flavor to the whisky in

distillery facts

- 1779
- Morrison-Bowmore Distillers Ltd.
- Islay Campbell
- River Laggan
- 2 wash 2 spirit
- Ex-bourbon and sherry
- Mon.–Fri. 10:00 until last tour 3:30; £2 entrance charge redeemable in the distillery shop

the barrels. Bowmore was built in 1779 and is one of the earliest recorded distilleries in Scotland. During World War II Bowmore was a base for Coastal Command flying boats. The distillery was purchased by Stanley P. Morrison in 1963. One of the bonds has been converted into a swimming pool for the local community and is heated by waste heat produced by the distillery.

Bowmore produces a wide range of truly distinctive malts from its own in-house maltings (it is one of a small number of distilleries to malt their own barley) which are dried over peat fired kilns, and matured in a mix of bourbon and sherry casks. The malts range in color from light gold to amber and bronze, reminiscent of the colors of a Bowmore sunset.

ages, bottlings, awards
Bowmore is currently bottled at unaged (Legend), 12, 17, 21, 25, and 30 years.
A rarity is Black Bowmore
Also special editions for the export market
1992 IWSC Best Single Malt Trophy 21 years
1994 Best Special Edition Malt, Black Bowmore
1995 Distiller of the Year

tasting notes

AGE: Legend 40%

NOSE: Peat with the tang of the sea.

TASTE: Flavors of the sea and smoke with citrus and a fresh warming finish.

AGE: 12 years 43%

NOSE: A light, smoky nose with a stronger hint of the sea.

TASTE: The heather of the peat and the tang of the sea combine to give a round, satisfying taste with a long finish.

AGE: 17 years 43%

NOSE: The smoky aroma has taken on hints of ripe fruits and flowers.

TASTE: A complex malt full of honey, seaweed, toffee, and citrus flavors with a long, mellow finish.
A perfect after-dinner malt.

Bruichladdich

BRUICHLADDICH, ISLAY, ARGYLL PA49 7UN
TEL: +44 (0)1496 850221

BRUICHLADDICH DISTILLERY sits on the edge of Loch Indaal and is Scotland's westernmost distillery. It was built in 1881 by Robert, William, and John Gourlay Harvey and was one of the first to be built using concrete. In 1886 the company was relaunched under the name of Bruichladdich Distillery Co. (Islay) Ltd. and continued producing whisky until 1929 when it was silent for some eight years. It is now part of Whyte & Mackay's malt whisky portfolio and is unfortunately silent again; it was mothballed in 1995.

Bruichladdich is a smoother single malt than the traditionally peaty Islay malts.

distillery facts

- 1881
- The Whyte & Mackay Group Plc.
- Non-operational
- Private reservoir
- 2 wash 2 spirit
- American white oak
- No visitors

tasting notes

AGE: 10 years 40%

NOSE: A refreshing subtle aroma.

TASTE: Medium-bodied with a lingering flavor, undertones of citrus and peat—a light Islay malt.

Bunnahabhain

ISLAY

BUNNAHABHAIN DISTILLERY, PORT ASKAIG, ISLE OF ISLAY,
ARGYLL PA46 7RR
TEL: +44 (0)1496 840646 FAX: +44 (0)1496 840248

DISTILLING HAS been part of life in Islay for more than 400 years. Bunnahabhain was built in 1883 to satisfy the blenders' demand for fine malt whiskies and in particular those from Islay. The site was chosen because of its accessibility by boat from the mainland and a ready supply of fresh peaty water. Bunnahabhain is Gaelic for "mouth of the river" and refers to the River Margadale at whose mouth the Greenless brothers built the distillery.

The distillery was built from local stone and the buildings formed a square with a central gateway. Building work included the construction of a mile-long road to join the route from Port Askaig, a pier, and houses for the workforce and the exciseman. The Islay weather often delayed work and on one occasion a gale damaged much of the structure and swept two new steam boilers off the beach and across the Sound of Islay onto Jura. Visitors can rent four cottages next to the distillery.

distillery facts

- 1883
- The Highland Distilleries Co. Ltd.
- Hamish Proctor
- River Margadale
- 2 wash 2 spirit
- Mix of bourbon and sherry
- By appointment only

"*Westering Home*"...

Bunnahabhain
SINGLE ISLAY MALT SCOTCH WHISKY
PRODUCT OF SCOTLAND
THE BUNNAHABHAIN DISTILLERY COMPANY.
BUNNAHABHAIN. ISLE OF ISLAY. SCOTLAND. BOTTLED IN SCOTLAND.

40% vol.   70 cl

tasting notes

AGE: 12 years 40%

NOSE: Definite aroma of sea and summer flowers.

TASTE: A surprise for an Islay malt with just a hint of peat, light and malty.
A richer, stronger finish.
A favorite with overseas drinkers as an after-dinner drink.

ages, bottlings, awards
Bunnahabhain 12 years 40%
Special 1963 Distillation

Initially, sales were made solely to the wholesale market for blended whisky. In the late 1970s the owners, Highland Distilleries, launched a 12-year-old Bunnahabhain, a lightly peated malt with a soft, mellow character and a golden corn color.

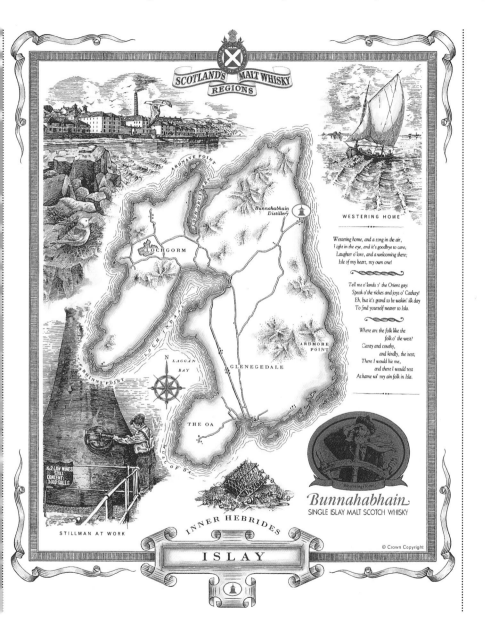

SCOTLAND'S MALT WHISKY
REGIONS

WESTERING HOME

Westering home, and a song in the air,
Light in the eye, and it's goodbye to care;
Laughter o' love, and a welcoming there;
Isle of my heart, my own one!

Tell me o' lands o' the Orient gay:
Speak o' the riches and joys o' Cathay!
Eh, but it's grand to be wakin' ilk day
To find yourself nearer to Isla.

Where are the folk like the
folk o' the west?
Canty and couthy,
and kindly, the best;
There I would hie me,
and there I would rest
At hame wi' my ain folk in Isla.

ARDNAVE POINT

Bunnahabhain
Distillery

LOCHGORM

ARDMORE
POINT

LOCH INDAAL

RHINNS POINT

N

LAGGAN
BAY

GLENEGEDALE

THE OA

MULL OF OA

No.2 LOW WINES
STILL
CONTENT
3,000 GALLS.

STILLMAN AT WORK

Bunnahabhain
SINGLE ISLAY MALT SCOTCH WHISKY

INNER HEBRIDES

ISLAY

Bushmills

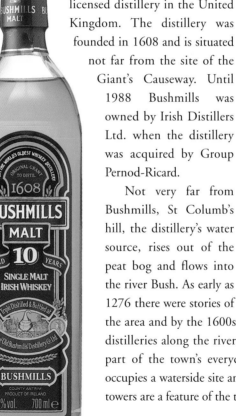

BUSHMILLS IS the oldest recorded licensed distillery in the United Kingdom. The distillery was founded in 1608 and is situated not far from the site of the Giant's Causeway. Until 1988 Bushmills was owned by Irish Distillers Ltd. when the distillery was acquired by Group Pernod-Ricard.

Not very far from Bushmills, St Columb's hill, the distillery's water source, rises out of the peat bog and flows into the river Bush. As early as 1276 there were stories of the spirit being produced in the area and by the 1600s there were many mills with distilleries along the riverside. Distilling had become part of the town's everyday life. The distillery still occupies a waterside site and its pagoda-shaped malting towers are a feature of the town.

distillery facts

- 1608
- Société Pernod-Ricard
- Frank McHardy
- St Columb's Rill
- 4 wash 5 spirit
- Ex-bourbon and sherry
- Mon.–Thurs. 9:00–12:00, 1:30–4:00
  In summer
  Fri. 9:00–4:00
  Sat. 10:00–4:00

As in Scotland, making whiskey in Ireland was a logical extension of a farmer's life and a means of using surplus grain. The copious supplies of pure water and peat for fuel ensured a steady supply of the raw ingredients required. At Bushmills, as at Auchentoshan and Rosebank in Scotland, the whiskey is triple-distilled. As a result the final spirit is simpler than whisky which is only distilled twice, since fewer constituent ingredients remain in the spirit. Knowing what makes any one whisky different from another is very difficult, but apart from triple distilling, it is true to say that Bushmills is slightly further south and therefore warmer, which could affect maturation in the barrels. There is no doubt that Bushmills is a very distinctive whiskey with a full flavor and a firm favorite.

ages and special bottlings
Bushmills 10 years 40%
Bushmills 16 and 19 years
Export 10 years 43%

tasting notes

AGE: 10 years 40%

NOSE: Warm, honey with sherry and spice.

TASTE: A warm, smooth malt with full flavors of sweetness and spice.
Recommended for after-dinner drinking.

Caol Ila

Caol Ila Distillery, Port Askaig, Islay, Argyll PA46 7RL
tel: +44 (0)1496 840207 fax: +44 (0)1496 840660

THE LABEL on a bottle of Caol Ila shows a drawing of a seal, since seals swim in the Sound of Jura in front of the distillery. Perhaps one of the finest views on the island is from the stillhouse looking out across the water to the mountains known as the Paps of Jura as they appear in and out of the mist.

The distillery was built in 1846 by Hector Henderson owner of the Camlachie distillery in Glasgow. Stone from the surrounding hills was used to build the distillery and houses for the workers. The distillery has its own jetty and supplies were brought in by puffer, and whisky taken away by steamers until comparatively recently. In 1927 Distillers Co. Ltd. obtained a controlling interest in Caol Ila and purchased their own puffer, the Pibroch. The distillery was largely rebuilt in 1974 and the number of stills increased to six from two. This is the largest distillery on the island and the modern buildings seem a little incongruous in the remote Islay countryside.

distillery facts

- 1846
- United Distillers
- Mike Nicolson
- Loch Nam Ban
- 3 wash 3 spirit
- N/A
- Please telephone for an appointment

RARE MALTS
SELECTION

NATURAL
CASK STRENGTH
SINGLE MALT
SCOTCH WHISKY
AGED **20** YEARS
DISTILLED 1975
CAOL ILA
DISTILLERY
ESTABLISHED 1846
PORT ASKAIG, ISLAY
61.12%vol 75cl
PRODUCED AND BOTTLED
IN SCOTLAND
LIMITED EDITION
BOTTLE Nº 0256

tasting notes

AGE: 15 years 43%

NOSE: Clean with the aromas of the sea, smoke, and apples.

TASTE: Caol Isla is distilled using 100% peated malt; however, this distinctive malt has a mellow flavor with a slight hint of the sea and a clean finish.
This is a malt to seek out.

AGE: 29 years Distilled 1975 61.12% Rare Malts Selection

NOSE: Stronger, peatier aroma.

TASTE: Mellow, dry, with a peaty, slightly salty taste.
A hint of sweetness and a long, smooth finish.

ages, bottlings, awards

Caol Ila 15 years 43%
from United Distillers
20 years distilled 1975 61.12%
limited edition United Distillers
Rare Malt Selection

RARE MALTS
S E L E C T I O N

Each individual vintage has been specially selected from Scotland's finest single malt stocks of rare or now silent distilleries. The limited bottlings of these scarce and unique whiskies are at natural cask strength for the enjoyment of the true connoisseur.

NATURAL
CASK STRENGTH
SINGLE MALT
SCOTCH WHISKY

AGED **20** YEARS

DISTILLED 1975

CAOL ILA
DISTILLERY
ESTABLISHED 1846
PORT ASKAIG, ISLAY

61.18%vol 75cl
PRODUCED AND BOTTLED IN SCOTLAND
LIMITED EDITION
BOTTLE N⁰ 14705

Caol Ila malt used to be available only from specialist retailers, but is now bottled by United Distillers and can be found in good liquor stores. Caol Isla is pale straw in color and has a well-rounded, slightly peaty taste. This is an extremely good introduction to Islay malts.

Caperdonich

CAPERDONICH DISTILLERY, ROTHES, MORAYSHIRE AB38 7BS
TEL: +44 (0)1542 783300

IN 1897 THE owner of Glen Grant distillery, Major James Grant, built a second distillery which was known for many years as Glen Grant No. 2. The single malt produced by this distillery had its own different characteristics. The distilleries were joined together by a pipe which crossed over the main road, and as one narrator of the time noted, "the Rothes streets flowed with whisky."

distillery facts

🗝 1897
📕 The Seagram Co. Ltd.
Willie Mearns
📇 The Caperdonich
〰 Burn
2 wash 2 spirit
🅰 N/A
🎞 N/A
ℹ By appointment only

SPEYSIDE

The distillery closed in 1902 and was rebuilt by Glenlivet Distillers Ltd. in 1965. The distillery was then renamed Caperdonich, after the Caperdonich well, the source of water for both distilleries. In 1967 the number of stills was increased from two to four. Caperdonich is not normally sold as single malt; it is used by Seagrams in their famous blends, but occasionally bottles can be obtained from specialist retailers.

Caperdonich is a pale, warm, gold single malt.

ages, bottlings, awards
Not bottled by the distillery, only available from specialists such as Gordon & McPhail

tasting notes
AGE: 1980 40%
NOSE: A soft aroma with hints of peat and sherry.
TASTE: Medium-bodied with a warm, fruity taste and a long smoky finish.

Cardhu

SPEYSIDE

CARDHU DISTILLERY, KNOCKANDO, ABERLOUR,
BANFFSHIRE AB38 7RY
TEL: +44 (0)1346 810204 FAX: +44 (0)1340 810491

JOHN CUMMING started farming at Cardow in Upper Knockando in 1813. Like many other farmers Cumming began distilling because the isolated location meant he could do so without drawing the attention of the exciseman. Officers used to stay at the farm, so Cumming's wife Ellen raised a red flag once they were safely eating dinner, to warn other local distillers. In 1824 John Cumming took out a license to distill whisky legally. The family continued to farm and distill whisky as tenants until Elizabeth Cumming, who had supervised the business for a total of 17 years on her own, purchased land next to the farm buildings and erected a new distillery. Cardow was purchased by John Walker & Sons Ltd. in 1893 and in 1925 the company merged with the Distillers Company Ltd. The distillery was rebuilt in 1960–61 and the number of stills increased from four to six.

distillery facts

- 1824
- United Distillers
- Charlie Smith
- Springs on the Mannoch Hill and Lyne Burn
- 3 wash 3 spirit
- N/A
- Jan.–Dec., Mon.–Fri. 9:30–4:30 May–Sept., Sat. also Coffee room, exhibition, and picnic area

ages, bottlings, awards
Cardhu 12 years 40%
1992 Toilet of the Year Award

In 1981 the name of Cardow was changed to Cardhu. There are 16 houses at Cardhu for employees and the farm still occupies 150 acres producing barley, sheep, and beef cattle. Cardhu is a golden amber single malt.

tasting notes
AGE: 12 years 40%
NOSE: Warm honey and spice—a hint of winter sunshine.
TASTE: Fresh on the palate, a hint of honey and nutmeg, with a smooth finish.

Clynelish & Brora

CLYNELISH, BRORA, SUTHERLAND KW9 6LB

TEL: +44 (0)1408 621444 FAX: +44 (0)1408 621131

CLYNELISH DISTILLERY was founded in 1819 by the Marquess of Stafford, who married the daughter of the Duke of Sutherland. The distillery was first licensed to James Harper. Records state: "The first farm beyond the people's lot (at Brora) is Clynelish, which has recently been let to Mr. Harper from the county of Midlothian. Upon this farm also there has just been erected a distillery at an expense of £750 [$1,200]." The next lessee was Andrew Ross and in 1846 George Lawson took over the lease. Lawson extended the distillery and replaced the stills. When the distillery was sold to Ainsle & Co., blenders in Leith, it was "a singularly valuable property" (*Harper's Weekly*, 1896). In 1912 fifty percent of Clynelish was acquired by The Distillers Company Limited.

distillery facts

- 1819, rebuilt 1967
- United Distillers
- Bob Robertson
- Clynemilton Burn
- 6 wash 6 spirit
- N/A
- Mar.–Oct. Mon.–Fri. 9:30–4:00 Nov.–Feb. by appointment only

ages, bottlings, awards

Clynelish is readily available at 14 years
from United Distillers
Clynelish 23 years old distilled 1972 57.1%
from United Distillers
Rare Malts Selection
Brora 1972 available from
Gordon & McPhail and 1982
from Cadenheads
ROSPA Gold Award for Safety

tasting notes

AGE: Clynelish 23 years, distilled 1972 57.1%

NOSE: Full of fruit and spice, warm, inviting.

TASTE: Smooth, slightly dry at first, growing with fruit
and sweetness, a strong, flavorful finish.
A rare malt to seek out.

A new distillery was built nearby in 1967–68. The
new distillery was given the name Clynelish and the old
distillery was closed for a while. When the distillery was
reopened it was renamed Brora in April 1975.

Brora and Clynelish have the same water source,
the Clynemilton Burn.

Cragganmore

CRAGGANMORE DISTILLERY, BALLINDALLOCH, BANFFSHIRE AB37 9AB

TEL: +44 (0)1807 500202 FAX: +44 (0)1807 500288

CRAGGANMORE DISTILLERY was founded in 1869 by John Smith. An experienced distiller, he managed the Macallan Distillery in the 1850s, founded Glenlivet in 1858, then managed Wishaw Distillery and finally went back to Speyside in 1865 as lessee of Glenfarclas. The distillery was built at Ayeon Farm near Strathspey Railway. John Smith died in 1886 and Cragganmore continued under the managership of his brother, George. The distillery was then managed by Gordon's son, John, who had learned the distillery trade in the Transvaal in South Africa. In 1923 Gordon Smith's widow sold the distillery to a group of businessmen. Cragganmore closed from 1941 to 1946. In 1964 the distillery was extended and the number of stills increased from two to four. Cragganmore became a member of The Distillers Company of Edinburgh in 1965.

Cragganmore is marketed by United Distillers as one of their "Classic Malt" range.

distillery facts

- 1869
- United Distillers
- Mike Gunn
- Craggan Burn
- 2 wash 2 spirit
- N/A
- Trade visitors only, by appointment

ages, bottlings, awards

Cragganmore 12 years 40%
from United Distillers
Cragganmore 1978 from Gordon &
MacPhail and 1982 from Cadenheads

tasting notes

AGE: 12 years 40%

NOSE: Dry, honey aroma.

TASTE: A pleasant medium-bodied
malt with a short smoky finish.

Craigellachie

CRAIGELLACHIE DISTILLERY, CRAIGELLACHIE, ABERLOUR,
BANFFSHIRE AB38 9ST
TEL: +44 (0)1340 881211 FAX: +44 (0)1340 881311

CRAIGELLACHIE DISTILLERY was built by Alexander Edward in 1891. The distillery sits on a hillside above the village of Craigellachie. It was purchased in 1916 by Sir Peter Mackie, the father of White Horse Blended Whisky. In 1927 Craigellachie was purchased by Distillers Company Ltd. The distillery was rebuilt in 1964 and the number of stills increased from two to four.

distillery facts

- 1891
- United Distillers
- Archie Ness
- Little Conval Hill
- 2 wash 2 spirit
- N/A
- No visitors

tasting notes

AGE: 22 years 60.2%

NOSE: Strong, full peaty aroma.

TASTE: Deceptively light, medium-bodied, smoky and spicy.

RARE MALTS
SELECTION

Each individual vintage has been specially selected from Scotland's finest single malt stocks of rare or now silent distilleries. The limited bottlings of these scarce and unique whiskies are at natural cask strength for the enjoyment of the true connoisseur.

NATURAL
CASK STRENGTH
SINGLE MALT
SCOTCH WHISKY

AGED 22 YEARS
DISTILLED 1973
CRAIGELLACHIE
DISTILLERY
ESTABLISHED 1888
CRAIGELLACHIE, BANFFSHIRE

60.2%vol 70cl e
PRODUCED AND BOTTLED
IN SCOTLAND
LIMITED EDITION
BOTTLE Nº 2220

**NATURAL
CASK STRENGTH**
SINGLE MALT
SCOTCH WHISKY

AGED **22** YEARS

DISTILLED 1973
CRAIGELLACHIE
DISTILLERY
ESTABLISHED 1888
CRAIGELLACHIE, BANFFSHIRE

PRODUCED AND BOTTLED
IN SCOTLAND
LIMITED EDITION
BOTTLE

ages, bottlings, awards
Craigellachie 14 years 43%
Craigellachie 22 years distilled 1973
60.2% limited edition
United Distillers Rare Malts Selection

Dailuaine

DAILUAINE DISTILLERY, CARRON, ABERLOUR, BANFFSHIRE AB38 7RE

TEL: +44 (0)1340 810361 FAX: +44 (0)1340 810510

LIKE MANY distilleries, Dailuaine started life as a farm. Dailuaine means "green vale" in Gaelic, and the distillery, founded by William Mackenzie in 1851, sits in a hollow next to the Carron Burn. In 1863 the Strathspey Railway connected with Carron on the other bank of the Spey. After Mackenzie's death, his wife Jane leased the distillery to a Mr. James Fleming of Aberlour; in 1879 her son Thomas became a partner in the business. Thomas Mackenzie continued to run the company, although he changed its name several times until his death at age 66 in 1915. As he had no heirs, the company was acquired by the Distillers Company Ltd.

Much of Dailuaine was destroyed by fire in 1917. The distillery was rebuilt soon after, and again in 1959–60. Until 1967 barley, coal, empty casks, and consignments of whisky were transported on the railway line to and from Dailuaine. In 1967 Beeching's axe fell (see p.224) and the Spey Valley railway was closed. The distillery locomotive Dailuaine No. 1, manufactured by Barclay of Kilmarnock in 1939, has been preserved by the company and can be seen running on the Strathspey Railway.

distillery facts

- 1851
- United Distillers
- Neil Gillies
- Ballieumullich Burn
- 3 wash 3 spirit
- N/A
- No visitors

ages, bottlings, awards

Dailuaine 16 years

Dailuaine 22 years distilled 1973

60.92% limited edition

United Distillers Rare Malts Selection

RARE MALTS
SELECTION

Each individual vintage has been specially selected from Scotland's finest single malt stocks of rare or now silent distilleries. The limited bottlings of these scarce and unique whiskies are at natural cask strength for the enjoyment of the true connoisseur.

NATURAL
CASK STRENGTH
SINGLE MALT
SCOTCH WHISKY

AGED **22** YEARS

DISTILLED 1973

DAILUAINE
DISTILLERY
ESTABLISHED 1851
CARRON, BANFFSHIRE

PRODUCED AND BOTTLED
IN SCOTLAND
LIMITED EDITION
BOTTLE

tasting notes

AGE: Dailuaine 22 years distilled 1973 60.92%

NOSE: A full smoky aroma with a hint of honey.

TASTE: Spicy, Christmas pudding on the tongue with a sweet, long, exhilarating finish.

Dallas Dhu

DALLAS DHU DISTILLERY, FORRES, MORAYSHIRE IV37 0RR

TEL: +44 (0)1309 676548

DALLUS DHU was built in 1898 by Wright & Greig, whisky blenders in Glasgow, in association with Alexander Edward. Dallas Dhu was designed by Charles Doig, a consulting engineer from Elgin who was responsible for many distilleries constructed in the malt whisky boom period of the late 1890s. This boom turned into a recession the following year. Many distilleries were forced to close and construction work was halted. Dallas Dhu continued to prosper, however, and in 1919 the company was purchased by J.R. O'Brien & Co. Ltd., distillers in Glasgow. In 1921 ownership changed again when the company was acquired by Benmore Distilleries Ltd. of Glasgow. Dallas Dhu was purchased by the Distillers Company Ltd. in 1929.

distillery facts

- 1898
- United Distillers
- Non-operational
- Altyre Burn
- N/A
- N/A
- Apr.–Sept. 9:30–6:30, Sun. 2:00–6:30. Oct.–Mar. 9:30–4:30, Sun. 2:00–6:30 Closed Thurs. afternoon and all day Fri.

Dallas Dhu was closed by United Distillers in 1983, and is now run by Historic Scotland as a "living museum." Stocks of Dallas Dhu have been bottled by United Distillers as part of their "Rare Malts" selection and are also available from specialist bottlers.

ages, bottlings, awards
24 years 59.9% from
United Distillers
12 years 40% from
Gordon & MacPhail

tasting notes
AGE: 12 years 40%
NOSE: Warm with sherry and peat.
TASTE: A well-rounded malt with smoke and a warm, slightly oak finish.

The Dalmore

DALMORE DISTILLERY, ALNESS, ROSS-SHIRE IV17 0UT
TEL: +44 (0)1349 882362 FAX: +44 (0)1349 883655

THE DALMORE means "the big meadowland" and takes its name from the vast grassland of the Black Isle, which lies opposite the distillery in the Firth of Cromarty. This is a spectacularly beautiful part of Scotland, and the visitor is rewarded with exceptional views. The road to Dalmore Distillery is narrow and leads down the wooded hillside to the distillery buildings which look out across Firth. The mudflats offer ornithologists a wealth of birds including waders and whooper swans which can be seen around the distillery outfall. Visitors are warned to be careful as the mud can be dangerous in certain places. The distillery was built at Ardross farm in 1839 by Alexander Matheson, a member of the Hong Kong trading company Jardine Matheson. He chose Ardross because it was close to the River Alness, was easily accessible by boat, and was in the heart of good barley country. The forest of Ardross covers the hills behind the distillery.

distillery facts

- 1839
- The Whyte & Mackay Group Plc
- Steve Tulevicz
- The River Alness
- 4 wash 4 spirit
- Mix of oloroso sherry and American white oak
- By appointment at 11:00 or 2:00 Mon., Tues., Thurs. early Sept.–mid June Tel. 01349 882362

Records show that in 1850 a Margaret Sutherland was a "sometime distiller." In 1886 the distillery was purchased by the Mackenzie family and in 1960 they joined forces with Whyte & Mackay Ltd. to form Dalmore-Whyte & Mackay Ltd. During World War I, production ceased at Dalmore. The American Navy, attracted by the deep waters of the Firth of Cromarty, used the distillery to manufacture mines. Saladin maltings were built at Dalmore in 1956 and in 1966 the number of stills was increased from four to eight.

Whisky at Dalmore matures in a mixture of American white oak and oloroso sherry casks. The slightly colder climate here in the north of Scotland may encourage slower maturation, and the flavor of the final product is influenced by the soft water, the slightly peaty malted barley, and the winds from the sea.

tasting notes

AGE: 12 years 40%

NOSE: A full, fruity aroma with hints of sherry sweetness.

TASTE: A good, full-bodied malt with overtones of honey and spice with a dry finish.

ages, bottlings, awards
The Dalmore 12 years 40%
Special Bottlings at 18, 21, and 30 years
Stillman's Dram exclusive range of ages
23, 27, and 30 years
Special Bottling of maximum 300
bottles at 50 years

THE DALMORE
DISTILLERY

INVERNESS

THE HIGHLANDS

ABERDEEN

GLASGOW

EDINBURGH

THE LOWLANDS

Dalwhinnie

DALWHINNIE DISTILLERY, DALWHINNIE, INVERNESS-SHIRE PH19 1AB

TEL: +44 (0)1528 522240

DALWHINNIE SINGLE malt is part of United Distillers' "Classic Malt" range.

Dalwhinnie distillery began operations as the Strathspey Distillery in 1898, and stands at a popular meeting place for drovers from the north and west. (Dalwhinnie is Gaelic for "meeting place.") The distillery is 1,073 feet above sea level and close to Lochan an Doire-uaine, a source of pure water which flows through peat countryside into the Allt an t-Sluie Burn. The first owners of Dalwhinnie distillery were not successful at first and the distillery was soon purchased by Mr. A. P. Blyth, who also owned a distillery in Leith, for his son. In 1905 the distillery was sold to Cook & Bernheimer of New York, prominent distillers in the United States, for $2,000. The distillery was acquired by Sir James Calder in 1920 and then in 1926 by the Distillers Company Ltd. After a fire in 1934, the distillery was closed and did not reopen until after World War II.

Dalwhinnie distillery is also the site of a meteorological office and the manager takes daily readings.

distillery facts

- 1898
- United Distillers
- Robert Christine
- Alt an t-Sluie Burn
- 1 wash 1 spirit
- N/A
- Easter–Oct.
  Mon.–Fri. 9:30–4:30
  Other times please
  call 01528 522268
  for an appointment

ages, bottlings, awards
Dalwhinnie 15 years 43%

tasting notes

AGE: 15 years 43%

NOSE: Dry, aromatic, summery.

TASTE: A beautiful malt with hints of honey and a lush, sweet finish.

Deanston

DEANSTON DISTILLERY, DEANSTON, NEAR DOUNE,
PERTHSHIRE FK16 6AG
TEL: +44 (0)1786 841422 FAX: +44 (0)1786 841439

DEANSTON DISTILLERY is unique, as it is housed inside an historic building, an old cotton mill which was designed by inventor Richard Arkwright. Cotton mills and whisky distilleries have a common requirement—a pure source of water. The distillery stands on the banks of the River Teith, which tumbles down from the Trossachs, a valley in central Scotland, and is a river renowned for its salmon and the purity of its water. The mill was originally water-driven and today has its own electricity power station. Both the main distillery building and maturation warehouses date from 1785. The building was converted into a full working distillery in 1966 and was purchased by Burn Stewart Distillers in 1990.

distillery facts

- 1966
- Burn Stewart Distillers Plc.
- Ian Macmillan
- River Teith
- 2 wash 2 spirit
- Refill and sherry
- No visitors

ages, bottlings, awards
Deanston 12, 17, and 25 years

Deanston is a pale gold malt with a smooth, mellow character. The 12-year-old includes a history of the Scottish Wars of Independence. The Deanston 25 years old is in a very distinctive oval-shaped bottled and only 2,000 are produced each year.

tasting notes

AGE: 12 years 40%

NOSE: A truly cereal aroma.

TASTE: The malt flavor hits the palate first and then citrus and honey notes come into play.

AGE: 17 years 40%

NOSE: At first dry and slightly peaty, with warmer aromas of sherry.

TASTE: A rich malt with sherry undertones and a peaty, dry finish.

AGE: 25 years 40%

NOSE: The longer maturation produces a fuller, sweeter malt with a rich aroma.

TASTE: The oak tannins hover around the mouth, while the overall taste is full-bodied and creamy with a smoky finish. A rare and exquisite malt.

Drumguish

DRUMGUISH DISTILLERY, GLEN YROMIE, KINGUSSIE,
INVERNESS-SHIRE PH21 1NS
TEL: +44 (0)1540 661060 FAX: +44 (0)1540 661959

THE STORY of Drumguish distillery is also the story of one family and, in particular, one man. The Christie family started to build Drumguish in 1962 close by the original distillery, which closed in 1911. Much of the work was carried out by George Christie himself and the building was completed in 1987. The new distillery produced its first spirit in December 1990.

Drumguish distillery is a hand-built stone building with an old, but operational, water wheel.

distillery facts

- 1990
- Speyside Distillery Co. Ltd.
- Richard Beattie
- River Tromie
- 1 wash 1 spirit
- N/A
- No visitors

tasting notes

AGE: Unaged 40%

NOSE: A light aroma with hints of honey and fruit.

TASTE: Soft on the palate with a little honey and a long, smooth finish.

ages, bottlings, awards
Bottled unaged 40%
Future bottlings will be made as they
become available
Special Christmas bottling of 100
bottles from first production after
three years' maturation

PRODUCT OF SCOTLAND

DRUMGUISH

SINGLE HIGHLAND MALT
SCOTCH WHISKY

DISTILLED, MATURED AND BOTTLED
IN SCOTLAND
DRUMGUISH DISTILLERY CO. LTD.
DRUMGUISH • INVERNESS-SHIRE

Dufftown

DUFFTOWN DISTILLERY, DUFFTOWN, KEITH, BANFFSHIRE AB55 4BR
TEL: +44 (0)1340 820224 FAX: +44 (0)1340 820060

THE DUFFTOWN-GLENLIVET Distillery Co. was founded in 1896 when the distillery was built inside an old meal mill. The original water wheel is still on site at Dufftown. In 1897 the distillery was taken over by P. Mackenzie & Co., owners of the Blair Athol distillery. In 1933 the company was purchased by Arthur Bell & Sons. The number of stills was increased from two to four in 1967 and again from four to six in 1979.

The label on a bottle of Dufftown features the kingfisher, a local bird with brilliant plumage that can be seen along the Dullan River, which flows past the distillery buildings.

distillery facts

🌱	1896
🏛	United Distillers
✍	Steve McGingle
〜	Jock's Well
🅰	3 wash 3 spirit
🛢	N/A
ℹ	No visitors

ages, bottlings, awards
Dufftown 15 years 43%

tasting notes

AGE: 15 years 43%

NOSE: Warm, fragrant.

TASTE: Smooth, slightly sweet, with a hint of fruit.
A delicious light malt.

HIGHLAND
SINGLE MALT *SCOTCH WHISKY*

DUFFTOWN

distillery was established near *Dufftown* at the end of the 19th The *bright flash* of the KINGFISHER can often be seen over the *DULLAN RIVER*, which flows past the *old stone buildings* of the *distillery* on its way *north* to the *SPEY*. This *single HIGHLAND MALT WHISKY* is typically *SPEYSIDE* in character with a *delicate, fragrant*, almost *flowery* aroma and taste which *lingers* on the *palate*.

AGED **15** YEARS

43% vol 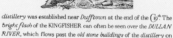 70 cl

The Edradour

EDRADOUR DISTILLERY, PITLOCHRY, PERTHSHIRE PH16 5JP
TEL: +44 (0)1796 473524 FAX: +44 (0)1796 472002

EDRADOUR DISTILLERY is Scotland's smallest distillery. Edradour, founded in 1825 on land rented from the Duke of Atholl, has changed little since then and is a good example of a working Victorian distillery. In 1886 the distillery was acquired by William Whiteley & Co. Ltd., a subsidiary of J. G. Turney & Sons of the United States. Edradour is owned today by Campbell Distillers, a part of the Pernod-Ricard.

The Edradour is a golden, honey-colored malt.

distillery facts

- 1825
- Campbell Distillers Ltd.
- John Reid
- Springs on Mhoulin Moor
- 1 wash 1 spirit
- N/A
- Mon.–Sat. 10:30–4:00 4:30–5:00 and Sun. 12:00–5:00

THE EDRADOUR
EST. 1825

ages, bottlings, awards
The Edradour 10 years 40%
and 43% export

tasting notes

AGE: 10 years 40%

NOSE: Delicate, sweet with a hint
        of peat.

TASTE: Dry, slightly sweet with a
        nutty, smooth finish.
        A malt for any occasion.

Glenallachie

GLENALLACHIE DISTILLERY, ABERLOUR, BANFFSHIRE AB38 9LR

TEL: +44 (0)1340 871315 FAX: +44 (0)1340 871711

GLENALLACHIE DISTILLERY was built in 1967 by W. Delme Evans for Mackinlay McPherson Ltd., a part of Scottish & Newcastle Breweries Ltd. Glenallachie nestles at the foot of Ben Rinnes mountain. In 1985 the distillery was purchased by Invergordon Distillers and in 1989 became part of Campbell Distillers. At present only stocks of the previous owners' bottlings are available.

distillery facts

- 1967
- Campbell Distillers Ltd.
- Robert Hay
- Springs on Ben Rinnes
- 2 wash 2 spirit
- N/A
- No visitors

tasting notes

AGE: 12 years 43%

NOSE: A light, flowery aroma.

TASTE: Delicate on the tongue with a hint of honey and fruit.
A long sweet finish.

ages, bottlings, awards

Available only from previous owners at 12 years

Glenburgie

GLENBURGIE DISTILLERY, BY ALVES, FORRES, MORAYSHIRE IV36 0QY
TEL: +44 (0)1343 850258 FAX: +44 (0)1343 850480

GLENBURGIE DISTILLERY was founded in 1810 as the Kilnflat Distillery and changed its name in 1871. In 1925 the distillery was managed by Margaret Nicol, who is believed to have been the first female manager ever. In 1936 Glenburgie was purchased by Hiram Walker and is now part of Allied Distillers' portfolio. The distillery is situated at the foot of Mill Buie Hills above the village of Kinloss and possesses attractive grounds.

A fine malt, Glenburgie is principally used in Ballantine blends. It is available at 18 years from Allied Distillers mainly for the export market and from Gordon & MacPhail at various ages.

distillery facts

- 1810
- Allied Distillers Ltd.
- Brian Thomas
- Local springs
- 2 wash 2 spirit
- Mixture of ex-bourbon and some sherry
- No visitors

tasting notes

AGE: 8 years 40%

NOSE: Scent of herbs and fruit.

TASTE: Strong at first with a lingering warm spice finish.

Glencadam

THE GLENCADAM DISTILLERY CO. LTD, BRECHIN, ANGUS DD9 6AY
TEL: +44 (0)1356 622217 FAX: +44 (0)1356 624926

IN 1825 George Cooper decided to license his distillery under the new Government Act legalizing whisky distilling and Glencadam was born. A history of Brechin relates that in 1838 there were two distilleries, two breweries, and 47 licensed premises. Today Glencadam is all that remains.

In 1891 Glencadam was purchased by Gilmour Thomson & Co. so they could ensure a consistent and good-quality supply of fine malts for their blends. During this time Gilmour Thomson's Royal Blend Scots Whisky enjoyed the patronage of HRH The Prince of Wales and had as its trademark the royal coat of arms and a stag.

distillery facts

- 1825
- Allied Distillers Ltd.
- Calcott Harper
- Loch Lee
- 1 wash 1 spirit
- Spanish oak
- 10:00–4:00 Mon.–Thurs.

tasting notes

AGE: 1974 40%

NOSE: A warm, sweet aroma with a touch of cinammon.

TASTE: A well-rounded malt with a hint of baked apples and cream and a warm finish.

Glen Deveron

MACDUFF DISTILLERY, BANFF, BANFFSHIRE AB4 3JT
TEL: +44 (0)1261 812612 FAX: +44 (0)1261 818083

CONFUSINGLY, MACDUFF distillery is the home of Glen Deveron single malt whisky. If you buy Glen Deveron malt from the independent bottlers, however, the label will show the contents as Macduff single malt whisky.

The distillery was founded on the banks of the River Deveron in 1962 by a consortium of businessmen and the company operated as Glen Deveron. Glen Deveron is now part of Bacardi Ltd.

Glen Deveron is a pale gold single malt.

distillery facts

- 1962
- Bacardi Ltd.
- Michael Roy
- Local spring
- 2 wash 3 spirit
- N/A
- No visitors

tasting notes

AGE: 12 years 40%

NOSE: Delicate, fresh.

TASTE: A medium-sweet malt with a long fresh finish.

The Glendronach

SPEYSIDE

GLENDRONACH DISTILLERY, FORGUE, HUNTLY,
ABERDEENSHIRE AB54 6DA
TEL: +44 (0)1466 730202 FAX: +44 (0)1466 730313

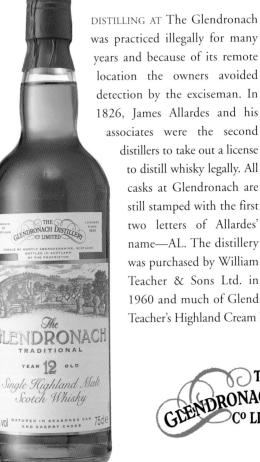

DISTILLING AT The Glendronach was practiced illegally for many years and because of its remote location the owners avoided detection by the exciseman. In 1826, James Allardes and his associates were the second distillers to take out a license to distill whisky legally. All casks at Glendronach are still stamped with the first two letters of Allardes' name—AL. The distillery was purchased by William Teacher & Sons Ltd. in 1960 and much of Glendronach malt is produced for Teacher's Highland Cream blended whisky.

distillery facts

- 1826
- Allied Distillers Ltd.
- Frank Massie
- Local springs
- 2 wash 2 spirit
- Seasoned oak and sherry
- Tours at 10:00 and 2:00
  Shop open during office hours

The Glendronach distillery sits just to the east of the Highland region of Scotland on the easternmost edge of the major whisky distilling area of Speyside. The malt is distinctive with some of the characteristics of a Speyside and others more closely associated with a Highland malt. In fact, some whisky writers place The Glendronach as being from the Highland region of Scotland.

Traveling to Huntly, the nearest town to the distillery from Aberdeen, the road winds through some spectacularly beautiful rolling countryside with a view of mountains always on the horizon. The area is well worth a visit, not just for the distillery, for it is here that some of Scotland's finest houses and castles are located. Visitors to Glendronach are rewarded with a view which has remained unchanged since the distillery's early days

of farming countryside, Highland cattle grazing alongside the road to the distillery, rooks nesting in the trees, and the manager's fine walled vegetable garden. At the time of this writing the distillery is mothballed, but there are sufficient stocks of The Glendronach widely available from wine merchants.

The Glendronach is a fine deep amber color and owes its character to the combination of in-house maltings, peat, and Highland water.

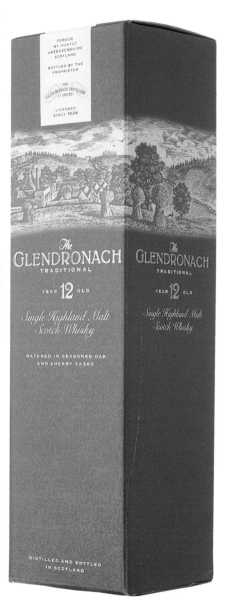

ages, bottlings, awards
The Glendronach 12 years 40% Traditional and 18 years Traditional
Glendronach 25 years 1968 matured in sherry casks
1993 *Decanter* magazine, highly recommended
1996 IWSC Silver Medal

tasting notes

AGE: 12 years 40% Traditional

NOSE: Sweet, smooth aroma.

TASTE: A long, sweet taste with smoky overtones and a pleasant finish.

Glendullan

GLENDULLAN DISTILLERY, DUFFTOWN, KEITH, BANFFSHIRE AB55 4DJ
TEL: +44 (0)1340 820250 FAX: +44 (0)1340 820064

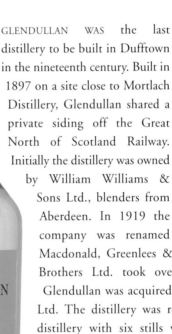

GLENDULLAN WAS the last distillery to be built in Dufftown in the nineteenth century. Built in 1897 on a site close to Mortlach Distillery, Glendullan shared a private siding off the Great North of Scotland Railway. Initially the distillery was owned by William Williams & Sons Ltd., blenders from Aberdeen. In 1919 the company was renamed Macdonald, Greenlees & Williams when Greenlees Brothers Ltd. took over the distillery. In 1926 Glendullan was acquired by the Distillers Company Ltd. The distillery was rebuilt in 1962 and a new distillery with six stills was added 1972. The old distillery was closed in 1985 and is used by United Distillers for maintenance workshops.

Glendullan distillery is still licensed to Macdonald, Greenlees Ltd. who are well-known whisky exporters. Their most famous brand is Old Parr blended whisky.

distillery facts

- 1897
- United Distillers
- Steve McGingle
- Springs in the Conval Hills
- 3 wash 3 spirit
- N/A
- No visitors

RARE MALTS
SELECTION

Each individual vintage has been specially selected from Scotland's
finest single malt stocks of rare or now silent distilleries.
The limited bottlings of these scarce and unique whiskies are at
natural cask strength for the enjoyment of the true connoisseur.

NATURAL
CASK STRENGTH
SINGLE MALT
SCOTCH WHISKY

AGED **23** YEARS

DISTILLED 1972
GLENDULLAN
DISTILLERY
ESTABLISHED 1897
DUFFTOWN, BANFFSHIRE

PRODUCED AND BOTTLED
IN SCOTLAND
LIMITED EDITION
BOTTLE

Estd. 1897
Glendullan
PURE *Highland* MALT

AGED **8** YEARS

*Distilled Slowly
and Matured for 8 long years in Oak Casks
for the Unique Flavour that is Glendullan*

40%vol   DISTILLED & BOTTLED BY
GLENDULLAN DISTILLERY, DUFFTOWN, BANFFSHIRE, SCOTLAND.   70cl e

ages, bottlings, awards

Glendullan 12 years 43%

Glendullan 22 years distilled

1972 62.6% limited edition

United Distillers Rare Malts Selection

tasting notes

AGE: 12 years 43%

NOSE: Delicate with a hint of almond.

TASTE: A warm honey malt with a long finish to savor.

Glen Elgin

Glen Elgin distillery, Longmor, Elgin, Morayshire IV30 3SL
TEL: +44 (0)1343 860212 FAX: +44 (0)1343 862077

GLEN ELGIN was designed by Charles Doig during the whisky boom of the 1890s. The boom ended when Pattisons of Leith, a firm of whisky-blenders, declared bankruptcy in 1899. The original owners had included William Simpson, who had been manager of Glenfarclas Distillery. Production at Glen Elgin began on May 1, 1900. The distillery was sold to the Glen Elgin-Glenlivet Distillery Co. Ltd. in 1901 and, for a while, production ceased. In 1906 the distillery was purchased by J. J. Blanche & Co. Ltd., wine-growers and shippers from Glasgow. However, production continued to be inconsistent. In 1930 Glen Elgin was purchased by the Distillers Company Ltd.

distillery facts

- 1898–1900
- United Distillers
- Harry Fox
- Local springs
- 4 wash 3 spirit
- N/A
- No visitors

tasting notes

AGE: no age 43%

NOSE: Smoky aroma with a hint of honey.

TASTE: Medium-bodied malt with a peaty taste, a hint of sweetness and a long finish.

Glenfarclas

SPEYSIDE

J AND G GRANT, GLENFARCLAS DISTILLERY, BALLINDALLOCH,
BANFFSHIRE AB37 9BD
TEL: +44 (0)1807 500245 FAX: +44 (0)1807 500234

A LICENSE was granted to Glenfarclas in 1836 just before Queen Victoria came to the throne. The distillery was built at Rechlerich Farm which is nestled at the foot of the Ben Rinnes Mountain in Speyside. In 1865 the lease passed to John Grant and the distillery quickly established itself as a favorite stopping off point for drovers on their way to market. The animals would stop for refreshment and so, too, would the men.

distillery facts

- 1836
- J. and G. Grant
- J. Miller
- spring on Ben Rinnes
- 3 wash 3 spirit
- Spanish oak
- 9:00–4:30 Mon.–Fri. year-round Sat. 10:00–4:00 June to Sept.

Glenfarclas is still owned by the same family and is a truly independent distillery. Many of the original distillery buildings have been modernized and the number of stills was increased from two to four in 1960 and again to six in 1976. The distillery has the largest stills and mash tun on Speyside.

The Glenfarclas visitor center has been fitted with original oak paneling from an old passenger liner, the *SS Empress* of Australia. Glenfarclas single malts are available at a wide variety of ages from ten to 30 years. They range in color from pale copper to glowing amber and often rate very highly on whisky connoisseurs' lists of favorite malts.

ages, bottlings, awards
Glenfarclas is bottled by the distillery at 10, 12, 15, 17, 21, 25, and 30 years at 40% and Glenfarclas 105 at cask strength 60%
1996 Wine & Spirit International Trophy Winner, Best Highland Single Malt Whisky—Glenfarclas 30-year-old

tasting notes

AGE: 105 60%

This cask-strength whisky is shown as unaged, but nothing is bottled at Glenfarclas until it is 10 years old.

This is the only malt readily available at this strength and has a warm golden color.

NOSE: This is a very pungent malt with a round, ripe aroma.

TASTE: On the tongue, a full sweet flavor with hints of caramel and a delicious aftertaste—not a malt for the faint-hearted.

AGE: 25 years 43%

NOSE: A warm aroma full of character and promise.

TASTE: The maturity of this single malt is apparent immediately, myriad flavors develop in the mouth and it has a long, slightly dry finish with oak undertones. A great malt.

Glenfiddich

WILLIAM GRANT & SONS LTD, THE GLENFIDDICH DISTILLERY,
DUFFTOWN, KEITH, BANFFSHIRE AB55 4DH

TEL: +44 (0)1340 820373 FAX: +44 (0)1340 820805

WILLIAM GRANT, the founder of William Grant & Sons, was determined to distill the "best dram [drink] in the valley." Glenfiddich was built by William Grant and his family—seven sons and two daughters—in 1886, and the first malt ran from the stills on Christmas Day in 1887. The distillery is still owned by William Grant's direct descendants, and they are completely committed to maintaining their independence and producing the finest whisky. Whisky is distilled following traditional methods; for example the distillery at Glenfiddich still has its own cooperage with nine coopers repairing and preparing barrels. The most modern part of the distillery is the highly automated bottling plant which produces 850,000 cases of Glenfiddich each year.

distillery facts

- 1886
- William Grant & Sons Ltd.
- Mr W. White
- Robbie Dubh
- 5 wash 8 spirit—unusually small
- All oak—American bourbon or Spanish sherry
- Year-round weekdays (except Christmas) 9:30–4:30
  Easter to mid-Oct.
  Sat. 9:30–4:30
  Sun. 12:00–4:30
  Groups of 12 or more please telephone beforehand

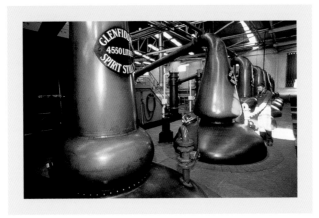

Glenfiddich malt is produced unaged, but is at least eight years old. The consistent Glenfiddich character is achieved by marrying casks together in large wooden marrying tuns for three to six months.

In 1963 Glenfiddich took the unusual step of marketing their whisky as a single malt in the United Kingdom and overseas. This was initially viewed with skepticism by other distillers, who continued to sell their malt to blenders but, thanks to the foresight of the Grant family, a market for single malts was created.

Glenfiddich Special Old Reserve single malt 40% is bottled in a very recognizable three-sided green bottle and is a pale golden color.

tasting notes

AGE: Unaged 40%

NOSE: A delicate, fresh aroma with a hint of peat.

TASTE: At first light, slightly dry then a fuller flavor develops with subtle, sweet overtones. A good all round malt, suitable for drinking throughout the day.

ages, bottlings, awards

Glenfiddich Special Old Reserve 40%
is bottled unaged (8 years minimum)
Glenfiddich Special Reserve
(vatting of 8–12 years casks)
Glenfiddich Excellence (18 years)
Glenfiddich Cask Strength (15 years)
1996 MPMA Gold Award for
The Glenfiddich Miniature Clan Tins

Glen Garioch

OLD MELDRUM, INVERURIE, ABERDEENSHIRE AB51 0ES
TEL: +44 (0)1651 872706 FAX: +44 (0)1651 872578

RECORDS SHOW that Glen Garioch was founded by Thomas Simpson in 1798. It is reputed, however, that Simpson was producing spirit in 1785, but whether at Glen Garioch or not is unclear. The Garioch (Geerie) valley is a very fertile stretch of Aberdeenshire and was the perfect place to build a distillery with a ready supply of barley. The distillery was purchased by several companies until it was closed in 1968. In 1970, Stanley P. Morrison (Agencies) Ltd. acquired the distillery and increased the number of stills.

Floor maltings were an important part of the distillation process at Glen Garioch and greenhouses were heated using waste heat. The distillery was mothballed in 1995.

Glen Garioch is available at various ages and ranges in color from pale gold to golden copper.

distillery facts

- 1798
- Morrison-Bowmore Distillers Ltd
- Ian Fyfe
- Springs on Percock Hill
- 2 wash 2 spirit
- Ex-bourbon and sherry Bottled strength varies with age
- No visitors

ages, bottlings, awards

Glen Garioch is bottled at unaged, 15, and 21 years

tasting notes

AGE: Unaged 40%

NOSE: Soft hint of peat and orange blossom.

TASTE: The first flavors in the mouth are peaty, then overtones of fruit and honey come into play with a long crisp finish.

AGE: 15 years 43%

NOSE: Warmer, fruitier aroma with hints of oak.

TASTE: A warm glowing whisky with citrus and smoke and a long mellow finish.

AGE: 21 years 43%

NOSE: Honey and peat with a slight hint of chocolate.

TASTE: Full-bodied, sweeter with a hint of smoke and a warm, mellow finish.

A good after-dinner malt.

Glengoyne

GLENGOYNE DISTILLERY, DUMGOYNE, STIRLINGSHIRE G63 9LB

TEL: +44 (0)1360 550229 FAX: +44 (0)1360 550094

A LICENSE was issued to Burnfoot Distillery in 1833 and leased to George Connell. In 1851–67 the distillery belonged to John McLelland and was then taken over by Archibald C. McLellan, who sold the distillery to Lang Brothers in 1876. The distillery was then renamed Glen Guin. The name was changed again in 1905 to Glengoyne. Glengoyne became part of Robertson & Baxter in 1965 and the distillery was rebuilt the year after with the addition of a new still for a total of three. Glengoyne is situated on the West Highland Way which makes a good stopping off point for ramblers as they make their way from Fort William to Glasgow.

distillery facts

- 1833
- Lang Brothers Ltd.
- Ian Taylor
- Burn from Campsie Hills
- 1 wash 2 spirit
- Ex-sherry and refill
- Mon.–Sat. 10:00–4:00 Sun. 12:00–4:00 Recommended by the Scottish Tourist Board

Glengoyne is a pale, white wine-colored malt made from unpeated barley.

ages, bottlings, awards
Glengoyne 10, 12, and 17 years 40%
Vintage bottlings
12 years 43% export

tasting notes

AGE: 10 years 40%

NOSE: A clean, sunny, floral aroma.

TASTE: Medium-bodied malt with hints of honey and a slight hint of fruit.
A good all-round malt whisky.

Glen Grant

SPEYSIDE

GLEN GRANT DISTILLERY, ROTHES, MORAYSHIRE AB38 7BS
TEL: +44 (0)1542 783318 FAX: +44 (0)1542 783306

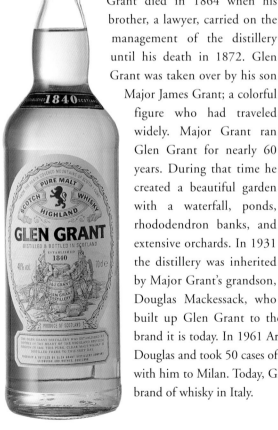

GLEN GRANT was founded by John and James Grant in 1840. John Grant died in 1864 when his brother, a lawyer, carried on the management of the distillery until his death in 1872. Glen Grant was taken over by his son Major James Grant; a colorful figure who had traveled widely. Major Grant ran Glen Grant for nearly 60 years. During that time he created a beautiful garden with a waterfall, ponds, rhododendron banks, and extensive orchards. In 1931 the distillery was inherited by Major Grant's grandson, Douglas Mackessack, who built up Glen Grant to the internationally renowned brand it is today. In 1961 Armando Giovinetti contacted Douglas and took 50 cases of Glen Grant 5 years old back with him to Milan. Today, Glen Grant is the number-one brand of whisky in Italy.

distillery facts

- 1840
- The Seagram Co. Ltd.
- Willie Mearns
- The Caperdonich Well
- 4 wash 4 spirit
- N/A
- Mid-Mar.–end Oct. Mon.–Sat. 10:00–4:00 and Sun. 11:30–4:00 Summer hours June to end Sept. Mon.–Sat. 10:00–5:00 and Sun. 11:30–5:00

ages, bottlings, awards
Glen Grant is sold unaged 40% for the
U.K. market
Export market 5 years

tasting notes

AGE: unaged 40%

NOSE: Dry, slightly tart.

TASTE: A light dry malt with a faint
hint of fruit in the finish.

Glen Keith

GLEN KEITH DISTILLERY, STATION ROAD, KEITH,
BANFFSHIRE AB55 3BU

TEL: +44 (0)1542 783042 FAX: +44 (0)1542 783056

ONE OF the first new distilleries to open in the twentieth century, Glen Keith was built on the site of a cornmill in 1958. The distillery is an attractive building constructed from local stone, located near the ruins of Milton Castle and a beautiful waterfall, the Linn of Keith.

Glen Keith was originally built with three stills for triple distillation. In 1970 the first gas-fired still in Scotland was installed at Glen Keith.

Glen Keith is used in fine blended whiskies including Passport. Visitors to Glen Keith can view an audiovisual tape entitled "Passport Experience," which tells the history of the blend. Glent Keith is marketed by Seagrams as part of their Heritage Selection range.

distillery facts

🖊	1958
	The Seagram Co. Ltd.
	Norman Green
〰	Balloch Hill springs
	3 wash 3 spirit
	N/A
ℹ	Please telephone for an appointment

tasting notes

AGE: 1983 43%

NOSE: Warm, scented with hints of oak and peat.

TASTE: A delicate malt with fruit and a hint of caramel and a long medium finish.

Glenkinchie

GLENKINCHIE DISTILLERY, PENTCAITLAND, EAST LOTHIAN EH34 5ET
TEL: +44 (0)1875 340333 FAX: +44 (0)1875 340854

GLENKINCHIE WAS founded in 1837 by John and George Rate, who had been operating the distillery under the name Milton from 1825 to 1833. Production ceased in 1853 and for some time the site was used as a sawmill. In 1880 the distillery was reinstated by a group of businessmen and in 1890 The Glenkinchie Distillery Co. Ltd. was formed. The distillery was completely rebuilt and the company continued operations until 1914, when it became part of Scottish Malt Distillers Ltd.

Glenkinchie ceased to malt its own barley in 1968. At that time, much equipment was being discarded by Glenkinchie and other distilleries in the group. The Museum of Malt Whisky Production was launched inside the old maltings facilities and includes a model of a Highland malt whisky distillery.

Glenkinchie distillery is located amid beautifully kept grounds and is open all year round to visitors. Glenkinchie is a pale malt.

distillery facts

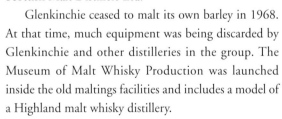 1837

United Distillers

Brian Bisset

Lammermuir Hills

I wash I spirit

No visitors

Mon.–Fri. 9:30–
4:00—Museum of
Malt Whisky
Production

ages, bottlings, awards

Glenkinchie 10 years 43% is bottled as part of United Distillers' Classic Malt range

tasting notes

AGE: 10 years 43%

NOSE: Orange blossom and honey.

TASTE: A smooth, light malt with a rounded flavor with a hint of sweetness and smoke and a long finish.
A malt for any time of day.

The Glenlivet

THE GLENLIVET DISTILLERY, BALLINDALLOCH,
BANFFSHIRE AB37 9DB
TEL: +44 (0)1542 783220

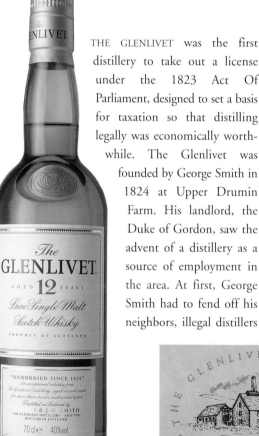

THE GLENLIVET was the first distillery to take out a license under the 1823 Act Of Parliament, designed to set a basis for taxation so that distilling legally was economically worthwhile. The Glenlivet was founded by George Smith in 1824 at Upper Drumin Farm. His landlord, the Duke of Gordon, saw the advent of a distillery as a source of employment in the area. At first, George Smith had to fend off his neighbors, illegal distillers

distillery facts

- 1824
- The Seagram Co. Ltd
- Jim Cryle
- Josie's Well
- 4 wash 4 spirit
- N/A
- Mid-Mar.–end Oct.
  Mon.–Sat. 10:00–
  4:00 Sun. 12:30–4:00
  July and Aug. open till
  6:00 daily
  Admission charge

who tried to burn the distillery down. However, with a pair of hair-trigger pistols he managed to persuade them to leave him alone. The Glenlivet Reception Center proudly displays these pistols. In 1858 George's son John joined him and they built a new distillery at Minmore Farm. The Glenlivet remained in the Smith family until 1975, when the owner, Captain Bill Smith, died. In 1977 the company was purchased by The Seagram Co. Ltd.

Many distilleries use the appellation "Glenlivet" but this is the only distillery which can call itself "The Glenlivet."

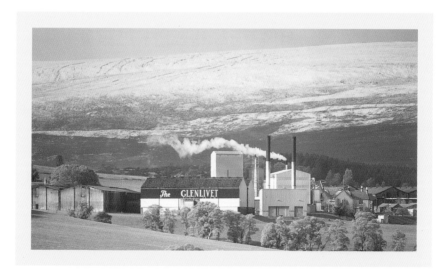

ages, bottlings, awards

The Glenlivet 12, 18, and 21 years

Only 1,000 bottles available of 18 years
in the United Kingdom

1995 IWSC Best Single Malt Scotch
Whisky Trophy for malts over 12
years—18 years The Glenlivet

tasting notes

AGE: 12 years 40%

NOSE: A fragrant malt with hints of
fruit.

TASTE: This is a beautiful malt with
a medium body and a
sweet, slightly sherry taste
with a long finish.

AGE: 18 years 43%

NOSE: A rich aroma with caramel
and peat.

TASTE: A gloriously rich malt, yet
dry, with fruit, peat, and a
spicy, sweet finish.
A fine, rare malt.

Glenlossie

GLENLOSSIE DISTILLERY, ELGIN, MORAYSHIRE IV30 3SS
TEL: +44 (0)1343 860331 FAX: +44 (0)1343 860302

GLENLOSSIE IS situated near Elgin, a town synonymous with whisky, and is adjacent to Mannochmore. Glenlossie Distillery was founded in 1876 by John Duff with John Hopkins, who ceased to work in 1888. A new company was formed, the Glenlossie-Glenlivet Distillery, which was then taken over by Scottish Malt Distillers in 1919. In 1962 the number of stills was increased from four to six. The spirit stills have purifiers just between the lyne arms and condensors, which add something different to this light fresh malt with a light lemon-gold color.

distillery facts

- 1876
- United Distillers
- Harry Fox
- The Bardon Burn
- 3 wash 3 spirit
- N/A
- No visitors

ages, bottlings, awards
Glenlossie 10 years 43%

tasting notes

AGE: 10 years 43%

NOSE: A light, fresh aroma with a delightful hint of honey and spice.

TASTE: Smooth with honey, smoke, and a little oak.

SPEYSIDE
SINGLE MALT *SCOTCH WHISKY*

The three *spirit stills* at the

GLENLOSSIE

distillery have *purifiers* installed between the *lyne arm* and the *condenser*. This has a bearing on the *character* of the *single MALT SCOTCH WHISKY* produced which has a *fresh, grassy* aroma and a *smooth*, lingering flavour. Built in 1876 by *John Duff*, the *distillery* lies four miles *south* of ELGIN in *Morayshire*.

AGED **10** YEARS

43% vol   Distilled & Bottled at SCOTZ-AND GLENLOSSIE DISTILLERY, Elgin, Moray, Scotland   70 cl

Glenmorangie

GLENMORANGIE DISTILLERY, TAIN, ROSS-SHIRE IV19 1PZ

TEL: +44 (0)1862 892043 FAX: +44 (0)1862 893862

THE DISTILLERY at Tain was first licensed in 1843 by William Mathieson and the first spirit was distilled in 1849. The building started out as a brewery run by McKenzie & Gallie. The first sales of Glenmorangie were made overseas in 1880—the *Inverness Advertiser* noted: "We observed the other day, en route for Rome, a cask of whisky from the Glenmorangie Distillery, likewise several casks destined for San Francisco." In 1887 the company was reconstructed as the Glenmorangie Distillery Co. Ltd. and in 1920 was purchased by Macdonald and Muir. The distillery was rebuilt in 1979 and the number of stills increased from two to four.

In 1996 the new Wood Finish Range was launched by Glenmorangie at a special presentation to celebrate the 80th birthday of former British Prime Minister Sir

distillery facts

- 1843
- Glenmorangie Plc.
- Bill Lumsden
- Tarlogie Springs
- 4 wash 4 spirit
- Old Madeira, port, or sherry casks
- Apr.–Oct. Mon.–Fri. 10:00–4:00
  Tours at 10:30 and 2:30
  Nov.–Mar. Mon.–Fri. 2:00–6:00
  Tours at 2:30 or please call 01862 892477 in advance
  Admission charge

Edward Heath. These special whiskies commemorate the fact that Macdonald and Muir traded in sherry, port, madeira, and claret right into the 1960s. These malts are well worth seeking out.

The malts vary in color from golden honey to golden amber to a beautiful copper with rose and gold.

tasting notes

AGE: Madeira Wood Finish,
12 years 43%

NOSE: Fresh, sweet, slightly nutty with citrus.

TASTE: Spicy with hints of citrus and honey with a dry finish.

AGE: Port Wood Finish,
12 years 43%

NOSE: Warm caramel yet fresh.

TASTE: Gloriously full and smooth on the mouth with hints of citrus and spice.

AGE: Sherry Wood Finish,
12 years 43%

NOSE: Sherry with malt and honey.

TASTE: Full-bodied with sherry and spice with a long, flavorful finish.

Glen Moray

GLEN MORAY DISTILLERY, ELGIN, MORAYSHIRE IV30 IYE
TEL: +44 (0)1343 542577 FAX: +44 (0)1343 546195

GLEN MORAY is located in the middle of one of the best farming areas of Scotland. The distillery began as a brewery and was converted in 1897 by the Glen Moray Glenlivet Distillery Co. Ltd. The site that Glen Moray stands on has an interesting bit of Scottish history attached to it: the old road into Elgin runs straight through the distillery in the lea of Gallow Crook, a site used for ex-ecutions until the end of the 1600s. The distillery closed in 1910 and was reopened by its present owners Macdonald and Muir Ltd. in 1923. At Glen Moray there is a sense of timelessness and the distillery still looks reminiscent of a Highland farm, constructed with buildings around a courtyard.

distillery facts

- 1897
- Glenmorangie Plc.
- Edwin Dodson
- River Lossie
- 2 wash 2 spirit
- N/A
- Please telephone for an appointment

tasting notes

AGE: Glen Moray 12 years 40%

NOSE: Delicate, hints of summer.

TASTE: A medium-bodied malt with a hint of peat and a warm, slightly sweet finish. A good after-dinner malt.

ages, bottlings, awards

Glen Moray 12 years 40% in a special blue tube Glen Moray 16 years in The Black Watch Highland Regiment tin

Glen Ord

HIGHLAND

GLEN ORD DISTILLERY, MUIR OF ORD, ROSS-SHIRE IV6 7UJ

TEL: +44 (0)1463 870421 FAX: +44 (0)1463 870101

ORD DISTILLERY was founded in 1838 by Robert Johnstone and Donald McLennan in an area renowned for whisky distilling. There were nine other small stills, all licensed. In 1860 Ord was acquired by Alexander McLennan, who was then declared bankrupt in 1871.

The distillery was passed to his widow, who eventually married again to Alexander McKenzie. He ran the business until 1887 when James Watson & Co., blenders of Dundee, purchased the distillery. Glen Ord became part of the Distillers Company in 1925.

Traditional maltings ceased in 1961 and a Saladin-box system was introduced. Much of the distillery was rebuilt in 1966.

distillery facts

- 1838
- United Distillers
- Kenny Gray
- Lochs Nan Eun and Nan Bonnach
- 3 wash 3 spirit
- N/A
- Mon.–Fri. 9:30–4:30

ages, bottlings, awards

Glen Ord 12 years 40%
ASVA Commended Exhibition
IWSC Best Single Malt up to 15 years
Grand Gold Medal at the
Monde Selection

tasting notes

AGE: 12 years 40%

NOSE: Full-bodied, warm, spicy.

TASTE: Flavorful with caramel,
nutmeg, and a long,
smooth finish.
Try Glen Ord in a cocktail.

Glenrothes

SPEYSIDE

GLENROTHES DISTILLERY, ROTHES, MORAYSHIRE AB38 7AA

TEL: +44 (0)1340 872300 FAX: +44 (0)1340 872172

THE GLENROTHES Distillery was built by W. Grant & Co. beside the Burn of Rothes, which flows from the Mannoch Hills, in 1878. Production started on Sunday December 28, 1879. In 1887 the Glenrothes distillery and the Islay Distillery Company, owners of Bunnahabhain Distillery, amalgamated to form the Highland Distilleries Co. Ltd. The water supply comes from The Lady's Well. Records show that this was where the only daughter of a fourteenth-century Earl of Rothes was murdered by the "Wolf of Babenoch" while attempting to save her lover's life.

In 1896, to meet increased demand, extension work was carried out at the distillery. In 1922, following a fire in a warehouse, whisky poured into the Burn of Rothes. Legend has it that the local population, and indeed a few cows too, took advantage of this free drink. In 1963 the number of stills was increased from four to six and again in 1980 from six to ten.

distillery facts

- 1878
- Highland Distilleries Co. Plc.
- A. B. Lawtie
- The Lady's Well
- 5 wash 5 spirit
- Varying mix of refills, sherry, and ex-bourbon
- By invitation only

THE
GLENROTHES
EST^D LIMITED RELEASE 1879
SINGLE SPEYSIDE MALT
Scotch Whisky

The Glenrothes Vintage is marketed by Berry Bros. & Rudd of London who market their own blend, Cutty Sark. Glenrothes is a fine single malt and has long been used by blenders.

The malts vary in color from very pale gold, through honey-gold to a rich amber.

tasting notes

AGE: 1972 43%

NOSE: Full caramel with spice.

TASTE: Full-bodied with warm oak
and honey notes.
A long, rich, sweet finish.

AGE: 1979 43%

NOSE: Warm caramel with faint
undertones of chocolate.

TASTE: Medium-bodied, flavorful
malt with hints of toffee and
orange in the mouth and a
long honey and citrus finish.

AGE: 1982 43%

NOSE: Warm caramel nose.

TASTE: Full-bodied with toffee and
vanilla and a long flavorful
finish.

AGE: 1984 43%

NOSE: Fine sherry, vanilla, and malt
aroma.

TASTE: Smooth, medium-bodied
with tropical fruit and malt
flavors and a long, smooth
finish.
A good after-dinner malt.

Glentauchers

GLENTAUCHERS DISTILLERY, MULBEN, KEITH AB55 6YL
TEL: +44 (0)1542 860272 FAX: +44 (0)1542 860327

THE FOUNDATION stone of the Glentauchers distillery was laid in May 1897 and distilling started the following year in 1898 by the Glentauchers Distillery Co., a partnership of three members of W. P. Lowrie & Co. Ltd., blenders, and James Buchanan & Co. Ltd. The distillery was largely rebuilt in 1965–66 and the number of stills was increased from two to six.

In 1985 Glentauchers was mothballed by United Distillers. The distillery was purchased by Allied Distillers in 1989. The distillery produces malts primarily for blended whiskies, and only a small amount of this iridescent gold single malt is available from specialist retailers.

distillery facts

- 1897
- Allied Distillers Ltd.
- William G. Wright
- From a dam fed by the Rosarie Burn
- 3 wash 3 spirit
- Mainly refill barrels
- No visitors

tasting notes

AGE: 1979 40%

NOSE: Fragrant, light, honeyed.

TASTE: A light, easy-going malt with a soft, dry finish.

Glenturret

GLENTURRET DISTILLERY, THE HOSH, CRIEFF, PERTHSHIRE PH7 4HA
TEL: +44 (0)1764 656565 FAX: +44 (0)1764 654366

BUILT IN 1775, Glenturret is Scotland's oldest highland malt whisky distillery. Research shows that distilling began in this area in 1717. Tracing the history of a distillery is often difficult and in the case of Glenturret records show that in the nineteenth century there were two distilleries in this area with the same name. However, the story of Glenturret by Hosh is well documented and by 1852 there was only one distillery operating. Glenturret is one of the smallest distilleries in Scotland. It is situated near the Turret Burn, which flows down from Turret Loch, a source of cool, clear water. Glenturret is now a subsidiary of Highland Distilleries.

Alfred Barnard wrote in 1887 that he "could see the chimney stack 120 feet high as he turned into the Glen, used in connection with the stills and boilers." Today more than 190,000 visitors each year take the road to the Glenturret distillery.

distillery facts

- 1775
- The Highland Distilleries Co. Plc.
- Neil Cameron
- Loch Turret
- 1 wash 1 spirit
- Mix of bourbon and sherry oak
- Mon.–Sat. 9:30–4:30, Jan.–Feb. Mon.–Fri. 11:30–2:30 Director of Tourism: Derek Brown

tasting notes

AGE: 12 years 40%

NOSE: Aromatic, hints of sherry and caramel.

TASTE: A full-bodied malt with wonderful warming flavor and a long, satisfying finish.

AGE: 15 years 40%

NOSE: Crisp, fresh yet sweet.

TASTE: Full flavor with angelica and spice with a long fruity finish.

ages, bottlings, awards
Glenturret 12, 15, 18, and 21 years—
Special Bottlings from time to time
Glenturret Malt Whisky Liqueur 35%
1974, 1981, 1991 IWSC Gold Medal
Le Monde Selection Brussels
Gold Medal 1990, 1991, 1994, and
many others

Highland Park

HIGHLAND PARK DISTILLERY, HOLM ROAD, KIRKWALL,
ORKNEY KW15 1SU
TEL: +44 (0)1856 873107 FAX: +44 (0)1856 876091

HIGHLAND PARK is the northernmost distillery in Scotland, situated on the Island of Orkney. The origins of the distillery are closely linked with the notorious smuggler Magnus Eunson whose story of evading the exciseman is told in this book (see p.11). The distillery was reputed to have been founded by David Robertson and passed through various hands until the stills, a malt barn, and other buildings were purchased by Robert Borwick in 1826. The distillery was built on a source of crystal-clear water emerging from two springs fed by a pool at Cattie Maggie's.

From 1826 the history of Highland Park is well-documented and in 1898 the number of stills was increased from two to four. The distillery was purchased by its current owners, Highland Distilleries, in 1935.

distillery facts

- 1798
- Highland Distilleries Co. Ltd.
- James Robertson
- Springs from Cattie Maggie's pool
- 2 wash 2 spirit
- Mix of sherry and bourbon oak casks
- Apr.–Oct. Mon.–Fri. 10:00–5:00 in July and Aug., also Sat. and Sun. 12:00–5:00 Nov., Dec., Mar. Mon.–Fri. 2:00 and 3:30 Closed Dec. 25 and 26, and all of Jan. and Feb.

An audio-visual presentation at Highland Park's visitor center shows images of Orkney interwoven with the production of whisky; Highland Park's own peat bogs, the standing stones of Maes Howe, the malting floors, and the waters of Scapa Flow which are now contained by the Churchill barrier built during World War II.

Highland Park single malt whisky is a glorious deep gold with a flavor that is enhanced by the distillery's own maltings over peat smoke and the salt-laden air.

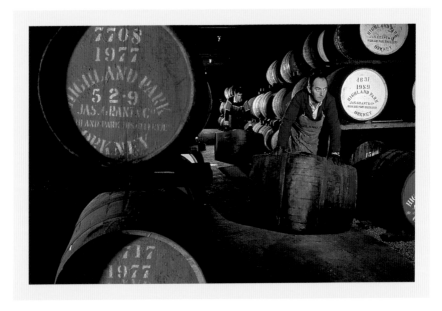

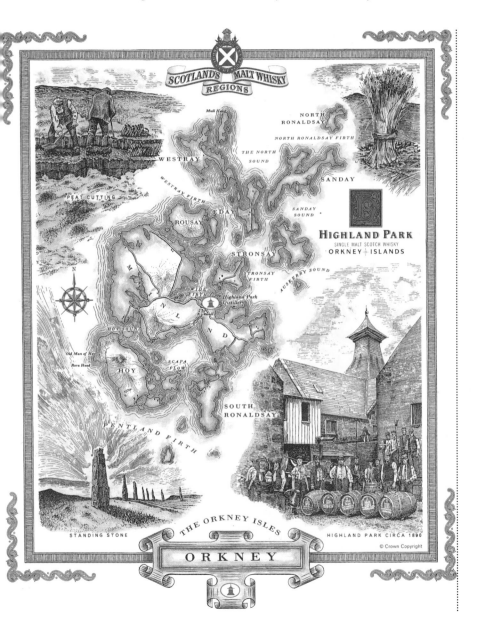

SCOTLAND'S MALT WHISKY REGIONS

MULL HEAD

NORTH RONALDSAY

NORTH RONALDSAY FIRTH

WESTRAY

THE NORTH SOUND

SANDAY

PEAT CUTTING

WESTRAY FIRTH

EDAY

SANDAY SOUND

ROUSAY

HIGHLAND PARK
SINGLE MALT SCOTCH WHISKY
ORKNEY ISLANDS

STRONSAY

STRONSAY FIRTH

AUSKERRY SOUND

N

M A I N L A N D

Highland Park Distillery

Kirkwall

HOY SOUND

Old Man of Hoy

Bera Head

HOY

SCAPA FLOW

SOUTH RONALDSAY

PENTLAND FIRTH

THE ORKNEY ISLES

STANDING STONE

HIGHLAND PARK CIRCA 1890

© Crown Copyright

ORKNEY

ages, bottlings, awards
Highland Park 12 years 40% from
Highland Distilleries
8 years 40% and 57% and cask strength
1984 from Gordon & MacPhail

tasting notes

AGE: 12 years 40%

NOSE: Rich, smoky with a hint
of honey.

TASTE: A glorious rounded malt
with heathery, peaty, and
warm, nutty overtones.
A dry yet sweet aftertaste.
A perfectly delicious after-
dinner malt.

Imperial

IMPERIAL DISTILLERY, CARRON BY ABERLOUR, BANFFSHIRE AB43 9QP

TEL: +44 (0)1340 810276 FAX: +44 (0)1340 810563

IMPERIAL DISTILLERY was built in 1897 by Thomas Mackenzie, then was transferred to Dailuaine Talisker Distilleries Ltd. in 1898. It occupies a site on the banks of the River Spey about three miles southwest of Aberlour.

Following a checkered history, the distillery was upgraded in the 1960s and operated Saladin maltings until 1984. In 1985 the distillery closed down, but has been in full production since 1989, when Allied Distillers purchased Imperial.

Imperial is a traditional Highland malt, which is much prized by connoisseurs and blenders alike. Very little Imperial is available as a single malt.

distillery facts

- 1897
- Allied Distillers Ltd.
- R. S. MacDonald
- The Ballintomb Burn
- 2 wash 2 spirit
- Ex-bourbon
- Please telephone for an appointment

tasting notes

AGE: 1979 40%

NOSE: Delightful aroma, flowers and smoke.

TASTE: A superb malt, sweet, and mellow with no harshness. Smooth, delicious aftertaste.

This single malt is not always readily available.

Inchgower

INCHGOWER DISTILLERY, BUCKIE, BANFFSHIRE AB56 2AB
TEL: +44 (0)1542 831161 FAX: +44 (0)1542 834531

INCHGOWER DISTILLERY was built in 1872 by Alexander Wilson to replace Tochineal distillery, which had been founded in 1832, also by Wilson. The Tochineal Distillery buildings still survive. Inchgower went into liquidation in 1930 and in 1936 Buckie Council purchased the distillery for $1,600. Arthur Bell & Sons Ltd. bought the distillery in 1938, and the number of stills was increased from two to four in 1966. Inchgower is now part of Distillers Company Ltd. Buckie is situated near the mouth of the River Spey and a bottle of Inchgower has an oyster-catcher on the label.

distillery facts

- 1872
- United Distillers
- Douglas Cameron
- Springs in the Menduff Hills
- 2 wash 2 spirit
- N/A
- No visitors

tasting notes

AGE: 14 years 43%

NOSE: Sweet, with a hint of apple.

TASTE: Medium-bodied malt with spice and a light, sweet finish.

ages, bottlings, awards

Inchgower 14 years 43%

Inchmurrin

LOCH LOMOND DISTILLERY, ALEXANDRIA, DUMBARTONSHIRE G83 0TL

TEL: +44 (0)1389 752781 FAX: +44 (0)1389 757977

INCHMURRIN DISTILLERY was founded in 1966 by the Littlemill Distillery Co. Ltd., which was a joint venture between Duncan Thomas and Barton Brands of the United States. Two types of single malt were produced by Littlemill—Inchmurrin and Rosdhu. In 1971 Barton Brands took over the company and built new blending and bottling facilities, although in 1984 the distillery closed. Inchmurrin reopened in 1987.

Inchmurrin is a very pale single malt.

distillery facts

- 1966
- Loch Lomond Distillery Co. Ltd.
- J. Peterson
- Loch Lomond
- 2 wash 2 spirit
- N/A
- No visitors

tasting notes

AGE: 10 years 40%

NOSE: Malty, spicy.

TASTE: A light-bodied spicy malt with a hint of lemon and a short finish.

Isle of Jura

ISLE OF JURA DISTILLERY, CRAIGHOUSE, JURA,
ARGYLLSHIRE PA60 7XT
TEL: +44 (0)1496 820240 FAX: +44 (0)1496 820344

ON THE west coast of Scotland, across the Sound of Islay, the mountain peaks of the Paps of Jura are a unique feature. Jura is one of the least populated Scottish islands with only some 200 inhabitants. The Isle of Jura Distillery is one of the main employers.

Jura's isolated situation encouraged illegal distilling and it is believed that whisky has been produced here since the late sixteenth century. The Isle of Jura Distillery was founded in 1810. There have been various owners, and it has frequently known periods of inactivity. The distillery was rebuilt in 1876, and again in the early 1960s, and now forms part of the Whyte & Mackay portfolio.

distillery facts

- 1810
- The Whyte & Mackay Group Plc.
- Willie Tait
- Market Loch
- 2 wash 2 spirit
- American white oak
- Please telephone for an appointment

tasting notes

AGE: 10 years 40%

NOSE: A golden malt with a peaty aroma.

TASTE: A light malt suitable for drinking as an aperitif yet with a full flavor and undertones of honey and smoke.

Knockando

KNOCKANDO DISTILLERY, KNOCKANDO, MORAYSHIRE AB38 7RT

TEL: +44 (0)1340 810205 FAX: +44 (0)1340 810369

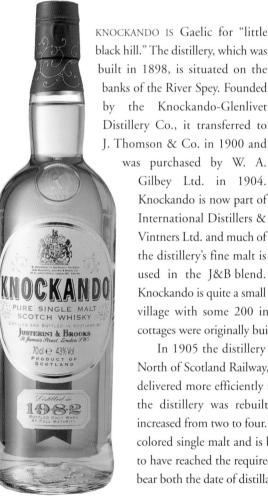

KNOCKANDO IS Gaelic for "little black hill." The distillery, which was built in 1898, is situated on the banks of the River Spey. Founded by the Knockando-Glenlivet Distillery Co., it transferred to J. Thomson & Co. in 1900 and was purchased by W. A. Gilbey Ltd. in 1904. Knockando is now part of International Distillers & Vintners Ltd. and much of the distillery's fine malt is used in the J&B blend. Knockando is quite a small village with some 200 inhabitants and many of the cottages were originally built for the distillery workers.

In 1905 the distillery was connected to the Great North of Scotland Railway, so that Knockando could be delivered more efficiently throughout Britain. In 1969 the distillery was rebuilt and the number of stills increased from two to four. Knockando is a pure golden-colored single malt and is bottled when it is considered to have reached the required level of maturation. Bottles bear both the date of distillation and the date of bottling.

distillery facts

- 1898
- International Distillers & Vintners Ltd.
- Innes A. Shaw
- Cardnach Spring
- 2 wash 2 spirit
- Ex-bourbon and sherry
- Please call 01340 810205 for an appointment

Knockando, matured for at least 12 years
Knockando Special Selection, matured
for at least 15 years
Knockando Extra Old, matured for at
least 20 years

tasting notes

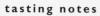

AGE: Distilled 1982
    Bottled 1996 43%
NOSE: Fragrant, spicy.
TASTE: A syrup-flavored malt with
    undertones of spice, vanilla,
    and filbert.

Lagavulin

ISLAY

LAGAVULIN DISTILLERY, PORT ELLEN, ISLAY, ARGYLL PA42 7DZ
TEL: +44 (0)1496 302400 FAX: +44 (0)1496 302321

ORIGINALLY, THERE were two distilleries at Lagavulin. The first was built in 1816 by John Johnston who continued producing whisky until 1833. The second was built by Archibald Campbell in 1817. In 1821 Campbell stopped distilling and both distilleries were then occupied by John Johnston from 1825 to 1834. However, as with all Islay distilleries, it is believed that illegal distilling started much earlier. Alfred Barnard, the whisky writer, wrote in 1887 that distilling was "the chief employment of the crofters and fishermen, more especially during the winter. In those days every smuggler could clear at least ten shillings a day, and keep a horse and cow."

In 1837 there was only one distillery owned by Donald Johnston. It was acquired by John Graham in 1852 and after various changes of ownership Lagavulin became part of the Distillers Company.

Lagavulin used small coasters to transport barley, coal,

distillery facts

- 1816
- United Distillers
- Mike Nicolson
- Solum Lochs
- 2 wash 2 spirit
- N/A
- Call 01496 302250 for an appointment

ages, bottlings, awards
Lagavulin 16 years 45% from
United Distillers as part of their
Classic Malt Range
1995, 1996 IWSC Gold Award

and empty casks from Glasgow, returning with full casks. These coasters were known as pibrochs and were used until the early 1970s when roll-on roll-off ferries were introduced to Islay.

Lagavulin Distillery is situated on a bay at Port Ellen with the ruins of Dunyveg Castle at its mouth.

tasting notes

AGE: 16 years 43%

NOSE: A very powerful,
       peaty smell.

TASTE: Full-bodied, pungent peat
       flavor with undertones of
       sweetness and a long finish.
       A perfect after-dinner malt.

Laphroaig

ISLAY

LAPHROAIG DISTILLERY, PORT ELLEN, ISLE OF ISLAY PA42 7DU
TEL: +44 (0)1496 302418 FAX: +44 (0)1496 302496

LAPHROAIG DISTILLERY was founded in 1815 by Alexander and Donald Johnston, who started farming at Laphroaig in around 1810. The first recorded distillery listing for Donald Johnston dates back to 1826. The distillery continued to remain in family ownership until 1908 when the proprietor, Ian Hunter, left the distillery to a Bessie Williamson. She was the first woman to run a malt whisky distillery in Scotland entirely on her own. Laphroaig has a distinctive flavor. During Prohibition it was legally imported to the United States because of its "medicinal" characteristics. Laphroaig is now part of the Allied Distillers malt whisky portfolio.

The distillery occupies an idyllic position close to the seashore with visiting otters and swans in residence. Laphroaig is one of the most popular single malts. An invitation to own a square foot of land adjacent to the distillery—is currently enclosed in bottles of Laphroaig. There are now thousands of proud owners.

distillery facts

- 1815
- Allied Distillers Ltd.
- Iain Henderson
- Kilbride Dam
- 3 wash 4 spirit
- American first-fill bourbon for single malt
- Call 01496 302418 for an appointment

Laphroaig is one of the few distillers to malt barley in-house, which is dried over fire kilns that burn local peat. Laphroaig is a vibrant golden malt.

tasting notes

AGE: 10 years 40%

NOSE: Instantly recognizable, full, peaty, slightly medicinal.

TASTE: A full-bodied malt with an initial peaty flavor, which develops to a touch of sweetness.

Long, dry, slightly salty finish.

ages, bottlings, awards

Laphroaig is bottled at 10 and 15 years

Vintage 1976 a total of 5,400 bottles some available through Duty Free

By appointment to HRH The Prince of Wales

1994 Queen's Award for Export Achievement

1993 IWSC Gold Medal—10 years

1985 IWSC Gold Medal—15 years

Linkwood

SPEYSIDE

LINKWOOD DISTILLERY, ELGIN, MORAYSHIRE IV30 3RD
TEL: +44 (0)1343 547004 FAX: +44 (0)1343 549449

LINKWOOD DISTILLERY was founded in 1825 by Peter Brown, agent for the Seafield Estates in Moray and Banffshire. His father farmed at Linkwood and it is likely that much of the barley used came from the farm and that the waste produced by the distillery was used as animal feed. In 1872 the distillery was rebuilt by his son, William Brown, and in 1897 the company was floated as the Linkwood-Glenlivet Distillery Co. Ltd., which was acquired by Scottish Malt Distillers. The number of stills at Linkwood was increased from two to six in 1971.

distillery facts

- 1825
- United Distillers
- Ian Millar
- Springs near Milbuies Loch
- 3 wash 3 spirit
- N/A
- By appointment only

RARE MALTS
SELECTION

Each individual vintage has been specially selected from Scotland's finest single malt stocks of rare or now silent distilleries. The limited bottlings of these scarce and unique whiskies are of natural cask strength for the enjoyment of the true connoisseur.

NATURAL
CASK STRENGTH
SINGLE MALT
SCOTCH WHISKY

AGED **23** YEARS

DISTILLED 1972

LINKWOOD
DISTILLERY
ESTABLISHED 1825
BY ELGIN MORAY

B297 **58.4**%vol 750ml
PRODUCED AND BOTTLED
IN SCOTLAND
LIMITED EDITION
BOTTLE Nº 0698

ages, bottlings, awards

Linkwood 12 years
Linkwood 20 years distilled 1972
58.4% limited edition United Distillers
Rare Malts Selection

tasting notes

AGE: 20 years distilled 1972
58.4%

NOSE: Full-bodied with fruit and
caramel.

TASTE: Full-bodied malt with honey
and a hint of peat and a
long, sweet finish.

SPEYSIDE
SINGLE MALT
SCOTCH WHISKY

LINKWOOD

distillery stands on the *River Lossie*,
close to *ELGIN* in *Speyside*. The *distillery*
has retained its *traditional atmosphere*
since its *establishment* in 1821.
Great care 🖎 has always
been taken to *safeguard* the
character of the *whisky* which has
remained the same through the
years. Linkwood is one of the
FINEST 🖎 *Single Malt Scotch Whiskies*
available – *full bodied* with a *hint* of
sweetness and a *slightly smoky aroma*.

YEARS **12** O L D

43% vol   Distilled & Bottled in *SCOTLAND*.
LINKWOOD DISTILLERY
Elgin, Moray, Scotland.   70 cl

Longmorn

LONGMORN DISTILLERY, NEAR ELGIN, MORAYSHIRE IV30 3SJ
TEL: +44 (0)1542 783400 FAX: +44 (0)1542 783404

LONGMORN DISTILLERY was built in 1894 by Charles Shirres, George Thomson, and John Duff. Power was provided by a large water wheel and the first distillation was produced in December 1894. In 1897 the Longmorn Distilleries Co. was launched, which owned both Benriach and Longmorn. In 1898, however, John Duff, who by then owned all the shares, ran into financial trouble. Hill, Thomson, & Co. Ltd. and the manager, James Grant, and his sons continued to run the distillery. The Grant family stayed in charge until 1970, when the company merged with The Glenlivet and Glen Grant Distillers Ltd. and traded stock as The Glenlivet Distillers Ltd. In 1978 the company was acquired by The Seagram Co. Ltd.

Longmorn is marketed by Seagrams as part of the "Heritage Selection" range and is widely available. Longmorn is a coppery-gold single malt.

distillery facts

- 1894
- The Seagram Co. Ltd.
- Bob MacPherson
- Local springs
- 4 wash 4 spirit
- N/A
- By appointment only

DISTILLED AND BOTTLED IN SCOTLAND

LONGMORN

Highland Single Malt

SCOTCH WHISKY

*This outstanding single malt whisky is produced only at the
Longmorn distillery, which stands on the site of an ancient abbey,
in the heart of the Scottish Highlands.*

MATURED IN OAK CASKS
15
YEARS

70 cl e   45% vol

ages, bottlings, awards
Longmorn 15 years 43% as part of
The Heritage Selection
1994 IWSC Gold Medal

tasting notes

AGE: 15 years 43%

NOSE: Fragrant, delicate, slightly
        fruity.

TASTE: Full of flavor with hints of
        fruit, flowers, and filberts
        with a long, sweet finish.

LONGMORN-DISTILLERY

The Macallan

THE MACALLAN DISTILLERY, CRAIGELLACHIE, BANFFSHIRE AB38 9RX
TEL: +44 (0)1340 871471 FAX: +44 (0)1340 871212

THE MACALLAN Distillery was founded in 1824 by Alexander Reid at the site of a ford across the River Spey at Easter Elchies. Easter Elchies Manor is now part of the distillery complex. After several changes of owner, the distillery was purchased by Roderick Kemp in 1892, who renamed it Macallan-Glenlivet. The distillery remained with the Kemp family until 1996 when it became part of the Highland Distilleries malt whisky portfolio.

distillery facts

- 1824
- Highland Distilleries Company Plc.
- David Robertson
- The Ringorm Burn
- 7 wash 14 spirit
- Ex-sherry oak Bottled strength varies according to age
- By appointment only

The number of stills was increased from six to 12 in 1965, in 1974 to 18, and finally in 1975 to 21. The stills are small, the spirit smaller than the wash, and all follow the original shape and size of the first stills. The Macallan is matured in old oak sherry casks, which impart a special flavor to the malt whisky. The Macallan ranges in color from palest gold to dark amber.

ages, bottlings, awards

The Macallan is bottled at 7, 10, 12, 18, and 25 years

Limited edition 60 years old with handpainted labels by Valerio Adami

Special 7-year-old bottling for Italy

Distillers Choice for Japan

Queens Award for Exports (twice)

IWSC Gold Medal 1996

PRODUCE OF   SCOTLAND

ESTABLISHED 1824

The

MACALLAN

Single Highland Malt Scotch Whisky

YEARS **10** OLD

DISTILLED AND BOTTLED BY
THE MACALLAN DISTILLERS LTD.
CRAIGELLACHIE · SCOTLAND

40% vol   BOTTLED IN SCOTLAND   70cl e

tasting notes

AGE: 10 years 40%

NOSE: Light, fragrant sherry.

TASTE: Full-bodied sherry with hints of vanilla and fruit, a long smooth well-rounded finish. A malt to savor before or after dinner.

Mannochmore

MANNOCHMORE DISTILLERY, ELGIN, MORAYSHIRE IV30 3SS

TEL: +44 (0)1343 860331 FAX: +44 (0)1343 860302

MANNOCHMORE 12-YEAR-OLD single malt has a drawing of a great spotted woodpecker on the label, an inhabitant of the Millbuies Woods which are next to the distillery. Mannochmore was founded in 1971 and was built alongside Glenlossie Distillery. Mannochmore was closed in 1985, but United Distillers reopened the distillery in 1989. Mothballed again in 1995, the distillery could yet be revived.

Mannochmore is a beautiful pale, gold malt.

distillery facts

- 1971
- United Distillers
- Non-operational
- The Bardon Burn
- 3 wash 3 spirit
- N/A
- No visitors

AGED **12** YEARS

tasting notes

AGE: 12 years 43%

NOSE: Delicate, springlike, with a hint of peat.

TASTE: A fine malt with a clean, fresh taste with a lingering, slightly sweet aftertaste.

SPEYSIDE
SINGLE MALT *SCOTCH WHISKY*

MANNOCHMORE

distillery stands a few miles *south* of Elgin in *Morayshire*. The nearby *Millbuies Woods* are rich in birdlife, including the Great *Spotted* Woodpecker. The *distillery* draws process *water* from the Bardon Burn, which has its *source* in the MANNOCH HILLS, and *cooling water* from the Gedloch Burn and the *Burn of Foths*. Mannochmore *single MALT WHISKY* has a *light, fruity* aroma and a *smooth,* mellow *taste*.

43% vol   AGED **12** YEARS   70cl

Miltonduff

MILTONDUFF DISTILLERY, MILTONDUFF, ELGIN, MORAYSHIRE IV30 3TQ
TEL: +44 (0)1343 547433 FAX: +44 (0)1343 548802

THE MILTONDUFF Distillery is situated in the Glen of Pluscarden on the site of the Pluscarden Priory on the banks of the Black Burn. Miltonduff was one of the first distilleries to take out a license in 1824.

History relates that there were more than 50 illegal distilleries on the same site and that smuggling continued to go on well into the nineteenth century. The Glen of Puscarden was an ideal spot for distilling, albeit illegally, as the surrounding hills form a triangle, which enabled the smugglers to devise a signaling system.

distillery facts

- 1824
- Allied Distillers Ltd.
- Stuart Pirie
- Black Burn
- 3 wash 3 spirit
- Usually ex-bourbon
- By appointment only Mon.–Thurs.

A flag would be hoisted onto one of the hills to warn of the approach of customs and excise officers. A conscientious exciseman learned of this practice, however, and hid all night until the men were in the fields. He arrived at the farmhouse to find the farmer's wife dismantling the still. She was apparently a very sturdy woman; he was much smaller. Legend has it that he was never seen again!

The distillery is now part of Allied Distillers and is the largest malt distillery in its portfolio. Most of the distillery's production goes into the blending of Ballantine's Whisky.

In the past a heavier malt was also produced at Miltonduff known as Mosstowie. The stills were removed in 1981. Supplies of Mosstowie single malt are available from the specialist bottlers Gordon & MacPhail.

ages, bottlings, awards

Miltonduff 12 years 43%

Also available at different ages from

Cadenheads of Edinburgh

tasting notes

AGE: 12 years 43%

NOSE: Fragrant.

TASTE: A medium-bodied malt with a fresh flavor.

Miyagikyo

SENDAI MIYAGIKYO DISTILLERY, NIKKA I-BANCHI, AOBA-KU,
SENDAI-SHI, MIYAGI-KEN 989034, JAPAN
TEL: +81 (0)22 395 2111 FAX: +81 (0)22 395 2861

THE HISTORY of the Nikka Whisky Distilling Co. Ltd. is fascinating. In 1918 Masataka Taketsuru, the son of a sake brewery owner, came to Glasgow University to learn about whisky. Masataka returned to Japan several years later with his Scottish bride, Jessie Rita. Using his new-found knowledge, he set about looking for the perfect site for a whisky distillery. The first was set up in Yoichi on Hokkaido Island, the most northern part of Japan, in 1934. The second distillery, Sendai, was built in 1969 to the north of the main island. Sendai is situated in a circle of mountains between two rivers. Sendai produces Miyagikyo, a deep mahogany-colored single malt.

distillery facts

- 1969
- Nikka Whisky Distilling Co. Ltd.
- Mr Yoshitomo Shibata
- Local springs
- 4 wash 4 spirit
- Mix of sherry, bourbon, refill, and new
- All year round with restaurant and shops

ages, bottlings, awards
Miyagikyo 12 years old 10,000 bottles
are produced each year, usually only
sold in Japan

tasting notes

AGE: 12 years

NOSE: Warm, sherry.

TASTE: Light with sherry, malt,
and vanilla.
A crisp finish.

Mortlach

MORTLACH DISTILLERY, DUFFTOWN, KEITH, BANFFSHIRE AB55 4AQ
TEL: +44 (0)1340 820318 FAX: +44 (0)1340 820019

MORTLACH DISTILLERY was founded in 1824 by James Findlater, Donald Mackintosh, and Alexander Gordon. In 1832 the distillery was acquired by A. & T. Gregory, who sold the distillery to J. & J. Grant of Glen Grant Distillery. The distillery was dismantled and did not operate again until 1842 when it was reinstated by John Gordon. In the early days the distillery was still a farm and waste barley was fed to the farm animals. George Cowie joined the company in 1854 and the distillery remained in his family's ownership until 1897 when Mortlach was purchased by John Walker & Sons Ltd. The distillery was completely rebuilt in 1963. The distillery became part of the Distillers Company in 1924.

distillery facts

- 1824
- United Distillers
- Steve McGringle
- Springs in the Conval Hills
- 3 wash 3 spirit
- N/A
- No visitors

ages, bottlings, awards
Mortlach 16 years 43%

tasting notes

AGE: 16 years 43%

NOSE: Fruity, warm with a hint of peat.

TASTE: Full-bodied with caramel and spice, and a long, sherry and honey finish.

SPEYSIDE
SINGLE MALT
SCOTCH WHISKY

MORTLACH

was the first of seven
distilleries in *Dufftown*. In the
19th *farm animals* kept in
adjoining byres were fed on
barley left over from processing.
Today *water* from springs in
the *CONVAL HILLS* is used to
produce this delightful
*smooth, fruity single
MALT SCOTCH WHISKY.*

AGED **16** YEARS

Distilled & Bottled in SCOTLAND.
MORTLACH DISTILLERY,
Dufftown, Keith, Banffshire, Scotland.

43% vol   70 cl

Oban

OBAN DISTILLERY, STAFFORD STREET, OBAN, ARGYLL PA34 5NH

TEL: +44 (0)1631 562110 FAX: +44 (0)1631 563344

OBAN DISTILLERY is one of United Distillers' Classic Malt range. Oban was founded in 1794 by the Stevensons, who were local businessmen with interests in quarrying, housebuilding, and shipbuilding. The distillery remained in the family's hands until 1866, when it was purchased by a local merchant, Peter Cumstie. Walter Higgin purchased Oban in 1883 and rebuilt the distillery. In 1898 Walter sold the distillery to the Oban and Aultmore-Glenlivet Distilleries Ltd. The directors included Alexander Edward, the owner of Aultmore and Messrs. Greig and Gillespie of whisky blenders Wright & Greig.

The buildings at Oban have remained virtually unchanged for nearly 100 years and are nestled right up against the cliffs, which rise 400 feet above the distillery.

distillery facts

- 1794
- United Distillers
- Ian Williams
- Loch Gleann a'Bhearraidh
- 1 wash 1 spirit
- N/A
- Year-round Mon.–Fri. 09.30–17.00 Also Sat. from Easter to Oct.

ages, bottlings, awards

Oban 14 years 43%

tasting notes

AGE: 14 years 43%

NOSE: Light with a hint of peat.

TASTE: Medium-bodied malt with a hint of smoke and a long, rewarding finish.

Old Fettercairn

FETTERCAIRN, DISTILLERY ROAD, LAURENCEKIRK,
KINCARDINESHIRE AB30 1YE
TEL: +44 (0)1561 340244 FAX: +44 (0)1561 340447

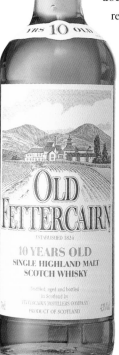

ALTHOUGH LEGEND tells of an earlier distillery in this area, some miles further up into the Grampian Mountains, no written documentation exists. However, records show that this distillery was built on this site at the foot of the hills in 1824 by Sir Alexander Ramsay. The building was originally a cornmill, which was destroyed by a fire in 1887. It was soon reconstructed and in 1966 the distillery was expanded increasing the number of stills from two to four. After several changes of ownership, Old Fettercairn was purchased by the Tomintoul Glenlivet Distillery Co. Ltd. in 1971 and is now part of the Whyte and Mackay Group Plc.

Old Fettercairn is a beautiful coppery-gold single malt whisky.

distillery facts

- 1824
- The Whyte & Mackay Group Plc.
- B. Kenny
- Springs in the Grampian Mountains
- 2 wash 2 spirit
- American white oak, oloroso sherry
- Mon.–Sat. 10:00–4:30 Call 01561 340205 to arrange group bookings

ages, bottlings, awards

Old Fettercairn 10 years 43%

tasting notes

AGE: 10 years 43%

NOSE: Delicate, fresh with a hint of smoke.

TASTE: A good introductory malt with a full flavor, undertones of peat, and a dry finish.

Rosebank

ROSEBANK DISTILLERY, CAMELON, FALKIRK,
STIRLINGSHIRE FK1 5BW
TEL: +44 (0)1324 623325

ALTHOUGH THE Rosebank distillery that survives today was reputed to have been built in 1840 in the maltings of the Camelon Distillery by James Rankine, records show that in 1817 there was another distillery of that name. The present distillery was rebuilt in 1864 and in 1894 renamed Rosebank. Rosebank has one wash and two spirit stills and is a triple-distilled malt. The distillery was mothballed in 1993. Rosebank is a summery, golden malt.

distillery facts

- 1840
- United Distillers
- Non-operational
- Carrow Valley reservoir
- 1 wash 2 spirit
- N/A
- No visitors

tasting notes

AGE: 1984 40%

NOSE: Fresh, with smoke and honey.

TASTE: Medium-bodied with a smooth, slightly dry citrus flavor.

Royal Brackla

ROYAL BRACKLA DISTILLERY, CAWDOR, NAIRN,
NAIRNSHIRE IV12 5QY
TEL: +44 (0)1667 404280 FAX: +44 (0)1667 404743

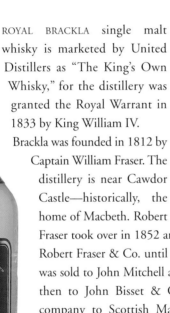

ROYAL BRACKLA single malt whisky is marketed by United Distillers as "The King's Own Whisky," for the distillery was granted the Royal Warrant in 1833 by King William IV.

Brackla was founded in 1812 by Captain William Fraser. The distillery is near Cawdor Castle—historically, the home of Macbeth. Robert Fraser took over in 1852 and the distillery continued as Robert Fraser & Co. until 1898. In 1898 the company was sold to John Mitchell and James Leict of Aberdeen, then to John Bisset & Co., who in turn sold the company to Scottish Malt Distillers in 1943. The distillery was rebuilt in 1965 and the number of stills was increased from two to four in 1970.

Royal Brackla is a fine golden malt.

distillery facts	
	1812
	United Distillers
	Chris Anderson
	The Cawdor Burn
	2 wash 2 spirit
	N/A
	No visitors

HIGHLAND
SINGLE MALT *SCOTCH WHISKY*

ROYAL BRACKLA

distillery, established in 1812, ℋ lies on the
southern shore of the MORAY FIRTH at *Cawdor* near *Nairn.*
Woods around the *distillery* are home to the *SISKIN;*
although a *shy bird,* it can often be seen *feeding* on *conifer* seeds.

In 1835 a *Royal Warrant* was granted to the *distillery* by King William IV,
who enjoyed the *fresh, grassy, fruity* aroma of this *single malt whisky.*

AGED **10** YEARS

43% vol   Distilled & Bottled at SCOTLAND ROYAL BRACKLA DISTILLERY, Cawdor Nairn, Scotland   70 cl

tasting notes

AGE: Unaged 40%

NOSE: Peat, honey, and spice.

TASTE: A medium-bodied malt with
a spicy sweetness and a
clean, slightly fruity finish.

Royal Lochnagar

ROYAL LOCHNAGAR, CRATHIE, BALLATER, ABERDEENSHIRE AB35 5TB
TEL: +44 (0)1339 742273 FAX: +44 (0)1339 742312

AS WITH many other distilleries, there were two Lochnagars. The first was built in 1826 and closed in 1860. In 1845 John Begg, a farmer, established the present distillery which was originally known as New Lochnagar. The distillery is situated close to Balmoral in the beautiful Deeside countryside and even today looks very much like a cluster of farm buildings. In 1848 John Begg wrote to Queen Victoria telling her that he believed the spirit was ready and extended an invitation to visit his distillery.

distillery facts

- 1845
- United Distillers
- Alastair Skakles
- Local springs
- 1 wash 1 spirit
- N/A
- Easter–Oct. Mon.– Sun. and Nov.–Easter, Mon.–Fri. 10:00–5:00. Multilingual display and restaurant

tasting notes

AGE: 12 years 40%

NOSE: Warm, spicy aroma.

TASTE: A whisky to savor, with fruit,
malt, and a hint of vanilla
and oak.
Sweet, long-lasting finish.

ages, bottlings, awards

Royal Lochnagar 12 years 40% and unaged

The Queen, Prince Albert, and their family arrived the very next day and thus Royal Lochnagar was born. HRH The Prince of Wales visited the distillery during a trip in 1996.

Royal Lochnagar is now part of the United Distillers malt whisky portfolio. The distillery plays an important part in the life of the local Crathie community with meetings and ceilidhs (traditional Scottish gatherings with music and recitations) held in the restaurant and in the converted barns. The distillery employees have also set up an award-winning nature reserve and, together with local schools, the flora and fauna of the area, even including the bat population, have all been well documented.

Scapa

SCAPA DISTILLERY, ST OLA, ORKNEY KW15 1SE
TEL: +44 (0)1856 872071 FAX: +44 (0)1856 876585

SCAPA IS one of Scotland's northern-most distilleries, situated on the banks of the Lingro Burn, and overlooking Scapa Flow on the Island of Orkney. It was built on the site of a former meal mill by J. T. Townsend. Scapa Distillery passed through many owners until the early 1950s when Hiram Walker of Allied Distillers purchased it from Bloch Bros. of Glasgow. The distillery was rebuilt in 1959 and continued production until 1993 when it was mothballed.

distillery facts	
🏷️	1885
🏭	Allied Distillers Ltd.
✍️	R. S. MacDonald
〰️	Springs
🅰️	1 wash 1 spirit
🛢️	Ex-bourbon
ℹ️	Please telephone for an appointment

The distillery benefits from a local supply of cool, clear water which rises in the ground to the north of the Orquil Farm.

During World War I the German Naval fleet sought refuge in Scapa Flow prior to a planned offensive. The assault never took place and the fleet was sunk deliberately by the German Navy. Relics of the ships can still be seen protruding above the waters of Scapa Flow.

A new Scapa 12-year-old single malt is bottled by Allied Distillers Ltd. Other ages can be obtained from Gordon & MacPhail.

ages, bottlings, awards
Scapa 12 years 40% Allied Distillers
Scapa 1985 available from
Gordon & MacPhail
1996 IWSC Gold Medal

tasting notes

AGE: 12 years 40%

NOSE: The Island of Orkney in a bottle—sea, peat, and heather.

TASTE: A mix of salt and citrus with a long-lasting, crisp finish. Try Scapa before dinner for a different cocktail.

The Singleton

SINGLETON AUCHROISK DISTILLERY, MULBEN, BANFFSHIRE AB55 3XS
TEL: +44 (0)1542 860333 FAX: +44 (0)1542
860265

THE SINGLETON Distillery is a relative newcomer to the malt whisky industry. Founded in 1974, it was first marketed in the United Kingdom as a single malt whisky in 1978. The distillery was opened by International Distillers and Vintners Ltd. and is now managed by its subsidiary, Justerini & Brooks (Scotland) Ltd. The distillery was built in traditional style and an old steam engine from Strathmill occupies a place of honor in the entrance hall. The distillery includes several warehouses for the maturation of various Highland and Speyside malts.

The Singleton has achieved worldwide acclaim as a single malt and is available at various ages. At 10 years old, 43%, The Singleton has a depth of color similar to beech leaves in the fall.

distillery facts

- 1974
- International Distillers & Vintners Ltd.
- Graeme Skinner
- Dorie's Well
- 4 wash 4 spirit
- Ex-bourbon and sherry
- Please telephone for an appointment

ages, bottlings, awards

The Singleton 10 years 43%
The Singleton Particular only available
in Japan matured for a minimum of
12 years
Many awards including:
1989 IWSC Best Malt Whisky
1992 & 1995 IWSC Gold Medal

tasting notes

AGE: 10 years 43%

NOSE: Rich, warming, with sherry notes.

TASTE: Full flavor on the tongue with hints of tangerine and honey, deliciously smooth in the mouth and a warm, long finish.
Recommended as an after-dinner malt.

Speyburn

SPEYBURN DISTILLERY, ROTHES, ABERLOUR, MORAYSHIRE AB38 7AG
TEL: +44 (0)1340 831231 FAX: +44 (0)1340 831678

SPEYBURN DISTILLERY was founded in 1897 by John Hopkins & Co., and is situated in a picturesque position nestled in the rolling hills of the Spey Valley. Tradition has it that distillation started before the building work had been completed and it was so cold that employees were forced to work in their overcoats.

The directors wanted very much to commemorate Queen Victoria's Diamond Jubilee on November 1 with their first fillings, but only one butt was bonded in 1897. Speyburn was one of the first malt distilleries to use pneumatic drum maltings, which were originally steam-driven.

distillery facts

- 1897
- Inver House Distillers Ltd.
- S. Robertson
- The Granty (or Birchfield) Burn, a tributary of the River Spey
- 1 wash 1 spirit
- Oak
- No visitors

ages, bottlings, awards
Speyburn 10 years 40% from
Inver House
Gold Medal and Named A Best Buy by
the Wine Enthusiast (United States)

tasting notes

AGE: 10 years 40%

NOSE: A dry, sweet-scented aroma

TASTE: A warm, flavorful malt
with hints of honey and a
herbal finish.
A malt to be savored
after dinner.

In 1916 Speyburn was acquired by the Distillers Company Ltd. and in 1992 was purchased by Inver House Distillers Ltd. Speyburn is a pale gold color reminiscent of larch trees in winter and is bottled by Inver House at 10 years old.

Springbank

J. & A. MITCHELL & CO. LTD, SPRINGBANK DISTILLERY,
CAMPBELTOWN, ARGYLL PA28 6EJ
TEL: +44 (0)1586 552085 FAX: +44 (0)1586 553215

SPRINGBANK DISTILLERY was built in Campbeltown on the Mull of Kintyre in 1828 by brothers Archibald and Hugh Mitchell on the site of their father's illegal still. It is believed that the Mitchells had been distilling for at least 100 years before this. By 1872 they owned four distilleries in the area. Over the years demand for Campbeltown whisky grew, as it was of a consistent quality and much prized by blenders. A supply of local coal to fire the stills encouraged the building of more and more distilleries. But unfortunately, with a few exceptions, distillers started to cut corners and, as the supply of coal started to run out, the quality deteriorated. By the 1920s the tide had turned and blenders were starting to look elsewhere for fine malts.

distillery facts

- 1828
- J & A Mitchell & Co. Ltd.
- Frank McHardy
- Crosshill Loch
- 2 wash 2 spirit
- Refill whisky, ex-bourbon, and sherry.
- By appointment only 2:00 weekdays June–September

The high standard of single malt whisky produced by the Mitchells never wavered, however, and today the distillery holds a prominent position in the center of Campbeltown. The present managing director is a direct descendant of Archibald Mitchell.

Production of a single malt at Springbank is entirely self-contained. From the very first stage of malting barley in the traditional way to the final step of bottling, everything is done on site. Springbank is one of only two distilleries to bottle at source—the other is Glenfiddich. The result of all this individual care is a highly prized malt with a rounded taste and rich aroma. Longrow has a heritage as old as Springbank. A fine malt of this name was produced in 1824 but the distillery, which was built next door to Springbank, was closed in 1896. The recipe for this peaty malt remained in the family and in 1973 a distillation was produced at Springbank which can be found from specialists. From time to time new distillations are produced and we understand that in 1997 a 10-year-old will be available.

ages, bottlings, awards

Springbank is available in 12, 15, 21, 25, and 30 years

1919 Springbank bottled in 1970—a very special 50-year-old single malt

tasting notes

AGE: 15 years 46%

NOSE: Fresh, rich with a hint of peat.

TASTE: Medium-bodied with an initial sweetness followed by a taste of the sea and oak. A long, smooth finish.

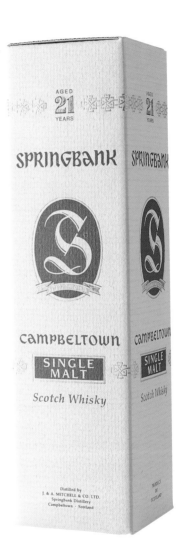

Strathisla

STRATHISLA DISTILLERY, SEAFIELD AVENUE, KEITH,
BANFFSHIRE AB55 3BS
TEL: +44 (0)1542 783042

STRATHISLA WAS founded by George Taylor and Alexander Milne in 1768 as the Milltown distillery. At that time the town of Keith was renowned for its linen mills. The distillery was then managed by several local businessmen and in 1830 was purchased by William Longmore. He was a local merchant and a key figure in Keith society, and the distillery grew under his control. In 1882 William Longmore died and the company was floated as William Longmore & Co. Ltd. Around this time the name was changed to Strathisla and major construction work was carried out, including installing a water wheel to drive the machinery and rebuilding the kiln with pagodas. In 1946 Longmore became a private company under the management of George

distillery facts

- 1786
- The Seagram Co. Ltd
- Norman Green
- Fons Bulliens' Well
- 2 wash 2 spirit
- N/A
- Feb. to mid-Mar. and Nov. Mon.–Fri. 9:30–4:00. Summer hours mid-Mar. to end Oct. Mon.–Sat. 9:30–4:00 and Sun. 12:30–4:00 Admission £4.00 including £2 voucher redeemable in the shop; free coffee and shortbread

Pomeroy, a London financier. He was found guilty of tax evasion and the company was closed in 1949. In 1950 it was sold to Chivas Brothers. Strathisla is one of the key malts in Chivas Regal, the premier Scotch whisky sold worldwide. Strathisla is marketed by Seagrams as part of their "Heritage Selection."

Strathisla is a warm coppery-gold single malt.

ages, bottlings, awards

Strathisla 12 years 43% in a distinctive flat bottle

tasting notes

AGE: 12 years 43%

NOSE: A beautiful aroma full of summer fruit and flowers.

TASTE: Light, sweet on the tongue with hints of peat and caramel. Long, smooth, fruity finish.

Talisker

TALISKER DISTILLERY, CARBOST, SKYE IV47 8SR

TEL: +44 (0)1478 640203 FAX: +44 (0)1478 640401

TALISKER IS the only distillery on the Isle of Skye and was built in 1830 on the edge of the Loch Harport by Hugh and Kenneth MacAskill. The MacAskills already had a reputation on the island, principally for driving the tenant farmers off the land, so that Hugh could then breed Cheviot sheep, and had acquired land around the Minginish peninsula, including Talisker House. Talisker is situated in a sheltered glen on the west coast of the island.

The story of Talisker is a checkered one and the distillery passed through many hands until it became part of the Distillers Company Ltd. in 1925. Talisker is marketed by United Distillers as one of their Classic Malt range.

distillery facts

- 1830
- United Distillers
- Mike Copland
- Cnoc-nan-Speireag
- 2 wash 2 spirit
- N/A
- Apr.–Oct. Mon.–Fri. 9:00–4:30
  July–Aug. also Saturdays 9:00–4:30
  Nov.–Mar. 2:00–4:30
  Large parties please telephone beforehand

ages, bottlings, awards
Talisker 10 years

ISLE OF SKYE
TALISKER
SINGLE MALT SCOTCH WHISKY

Beyond Carbost Village
close to the Shore is a
gentle haven sheltered
from the bleak ravines
which sweep down to
the coast.
Here in the shadow of the
distant Cuillin Hills lies
the islands only distillery
Talisker
The Golden Spirit of Skye
has more than a hint of
local seaweed peppered
with sour & sweet notes
and a memorable warm
peaty finish.

45.8% VOL TALISKER DISTILLERY CARBOST SKYE 70 cl e

tasting notes

AGE: 10 years

NOSE: Full, sweet yet peaty.

TASTE: A well-rounded, full-flavored malt with peat and honey and a lingering finish. A good all-round malt, try this before dinner.

Tamdhu

TAMDHU DISTILLERY, KNOCKANDO, ABERLOUR AB38 7RP
TEL: +44 (0)1340 870221 FAX: +44 (0)1340 810255

IN 1863 THE Highlands along the Upper Spey became more accessible to tourists with the opening of the Strathspey Railway from Boat of Gaten to Craigellachie. In the early 1890s blended whisky was increasing in popularity and businessmen were encouraged to invest in new distilleries. The new railway meant easier access to the Upper Spey, an area already renowned for its high-quality malts.

Tamdhu was built in 1896 by William Grant, a director of Highland Distilleries. The state-of-the-art distillery incorporated the latest devices, such as grain elevators and mechanical switchers for ease of production.

Like many distilleries, Tamdhu suffered during the war years and the Depression and closed from 1928 until after World War II. The distillery reopened and demand continued to grow until 1972 when the number of stills was increased from two to four and then again in 1975, when two more stills were added. In 1976 Tamdhu was launched as a single malt. Tamdhu is Gaelic for little dark hill.

distillery facts

- 1896
- Highland Distilleries Co. Ltd.
- W. Crilly
- Private springs
- 3 wash 3 spirit
- Mix of sherry and refill casks
- i No visitors

ages, bottlings, awards
Tamdhu is bottled unaged by
Highland Distilleries

tasting notes

AGE: Unaged 40%

NOSE: A light, warm aroma with a
hint of honey.

TASTE: Medium flavor, fresh on the
palate with overtones of
apple and pear orchards and
a long, mellow finish.
Suitable for drinking at
any time.

Tamdhu is the only Speyside distillery to malt all its own barley on site. Tamdhu single malt is the color of pale amber fall leaves and has a fresh and aromatic taste full of summer orchards and flowers.

Teaninch

TEANINCH DISTILLERY, ALNESS, ROSS-SHIRE IV17 0XB
TEL: +44 (0)1349 882461 FAX: +44 (0)1349 883864

TEANINCH STANDS on the River Alness not far from Cromarty Firth. The distillery was built by Captain Hugh Munro, owner of the Teaninch Estate in 1817. In the beginning it was difficult to obtain supplies of barley, as most went to the illicit distillers. However, by the 1830s production had increased by 30 times. Lieutenant General John Munro took over the distillery, but as he was away in India much of the time he granted a lease to Robert Pattison in 1850. In 1869 the lessee was John McGilchrist Ross who gave up in 1895. The distillery was taken over by Munro & Cameron of Elgin, who purchased it in 1898, and soon renovated and expanded it. In 1904 Innes Cameron became the sole owner who also had interests in Benrinnes, Linkwood, and Tamdhu. In 1933, a year after Innes Cameron died, his trustees sold Teaninch to Scottish Malt Distillers Ltd.

distillery facts

	1817
	United Distillers
	Angus Paul
	The Dairywell Spring
	3 wash 3 spirit
	N/A
i	No visitors

ages, bottlings, awards

Teaninch 10 years

Teaninch 23 years distilled 1972

64.95% limited edition

United Distillers Rare Malts Selection

RARE MALTS
SELECTION

Each individual vintage has been specially selected from Scotland's finest single malt stocks of rare or now silent distilleries. The limited bottlings of these scarce and unique whiskies are at natural cask strength for the enjoyment of the true connoisseur.

NATURAL
CASK STRENGTH
SINGLE MALT
SCOTCH WHISKY

AGED **23** YEARS

DISTILLED 1972

TEANINICH
DISTILLERY

ESTABLISHED 1817
ALNESS, ROSS-SHIRE

64.95% Alc/Vol (129.9 proof) 750ml

PRODUCED AND BOTTLED
IN SCOTLAND
LIMITED EDITION
BOTTLE Nº 6552

tasting notes

AGE: 23 years, distilled 1972
64.95%

NOSE: Light, peaty.

TASTE: Smoke and oak with a long, mellow finish.

Tobermory

TOBERMORY DISTILLERY, TOBERMORY, ISLE OF MULL,
ARGYLLSHIRE PA75 6NR
TEL: +44 (0)1688 302645 FAX: +44 (0)1688 302643

TOBERMORY HAS one of the most beautiful distillery locations at the southern end of the famous harbor on the Hebridean island of Mull. The only distillery on Mull, it was founded in 1795 by John Sinclair, a local merchant. The distillery became fully operational in 1823. Records show that in 1823 the land was granted to Sinclair by the British Society for Extending the Fisheries and Improving the Sea Coasts of the Kingdom. The distillery was silent between 1930 and 1972 and was then reopened as the Ledaig Distillery (Tobermory) Ltd. The company went into receivership in 1975 and was acquired by the Kirkleavington Property Co. of Cleckheaton, Yorkshire in 1978. The distillery closed again in 1989 and was then purchased four years later by Burn Stewart Distillers.

Tobermory is a pale straw-colored malt in a distinctive green bottle with white lettering.

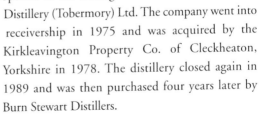

distillery facts

- 1795
- Burn Stewart Distillers Plc.
- Ian MacMillan Asst: Alan MacConochie
- Private loch
- 2 wash 2 spirit
- Refill
- Mon.–Fri. 10:00–4:00 Apr.–Sept.

Tobermory is made from unpeated barley. Some distillations are made using peated barley and these are bottled as Ledaig, although some older bottlings of Tobermory were also called Ledaig. All future bottlings of Ledaig will only be made using peated barley. Ledaig is a warm, golden-colored malt.

ages, bottlings, awards
Tobermory is bottled unaged 40%
Ledaig 1974 Vintage 43%
Ledaig 1975 Vintage 43%

tasting notes

AGE: Tobermory unaged 40%

NOSE: The island in a bottle—a light, soft, heathery aroma.

TASTE: Light, medium-flavored malt with undertones of honey and herbs and a soft, smoky finish.

AGE: Ledaig 1974 Vintage 43%

NOSE: Full-bodied with a strong, peaty aroma.

TASTE: Very flavorful in the mouth with peat and a hint of sherry.
Long, mellow finish.

Tomatin

TOMATIN DISTILLERY, TOMATIN, INVERNESS-SHIRE IV13 7YT
TEL: +44 (0)1808 511444 FAX: +44 (0)1808 511373

TOMATIN DISTILLERY, which stands at 1,028 feet above sea level, was founded in 1897. The company went into liquidation in 1906 and reopened again in 1909. In 1956 the number of stills was increased from two to four and further stills were steadily added until in 1974 there were a total of 23. One of the largest distilleries in Scotland, Tomatin is a subsidiary of Takara Shuzo & Okura of Japan and was the first distillery in Scotland to be purchased by a Japanese company.

distillery facts

- 1897
- Takara Shuzo & Okura & Co. Ltd
- T. R. McCulloch
- Allt na Frithe Burn
- 12 wash 11 spirit
- N/A
- Mon.–Fri. 9:00–4:30 Sat. May–Oct. 9:30–1:00 Please telephone for large groups and during Dec. & Jan.

ages and special bottlings

Tomatin 10 years 40%

Limited edition 25 years

Export 10 & 12 years

tasting notes

AGE: 10 years 40%

NOSE: A delicate aroma with hints of honey and smoke.

TASTE: Light and smooth with a hint of peat.

Tomintoul

TOMINTOUL, BALLINDALLOCH, BANFFSHIRE AB37 9AQ
TEL: +44 (0)1807 590274 FAX: +44 (0)1807 590342

TOMINTOUL IS a modern distillery built in 1964. Tomintoul is the second-highest village in Scotland, at 1,100 feet, and had a reputation in the past for illegal distilling. The distillery looks a little incongruous in this beautifully wooded, mountainous part of Scotland. The distillery was founded by a company called Hay & Macleod Ltd. and W. & S. Strong Ltd, Glasgow whisky brokers. It was acquired by Scottish & Universal Investment Trust in 1973 and now forms part of The Whyte & Mackay Group's malt whisky portfolio. The number of stills was increased from two to four in 1974.

Tomintoul has a warm, coppery-gold color.

distillery facts

- 1964
- The Whyte & Mackay Group Plc.
- R. Fleming
- Ballantruan Spring
- 2 wash 2 spirit
- N/A
- No visitors

ages, bottlings, awards
Tomintoul 12 years
Tomintoul 10 years 40% in the U.K.

tasting notes

AGE: 10 years 40%

NOSE: Light, sherry.

TASTE: Sweet on the tongue with smoky undertones.

The Tormore

SPEYSIDE

ALLIED DISTILLERS LTD, TORMORE DISTILLERY, ADVIE,
GRANTOWN-N-SPEY, MORAY PH26 3LR
TEL: +44 (0)1807 510244 FAX: +44 (0)1807 510352

TORMORE WAS the first new distillery to be built in the twentieth century in Scotland. Designed by Sir Alfred Richardson, it is an architectural gem. The distillery buildings and housing are built around a square with a belfry, which has a chiming clock.

Tormore, in the heart of the country, features landscaped gardens with an ornamental lake, fountains, and a topiary with a backdrop of pine-clad hills. The distillery is situated on the south side of the A95 road between Grantown-on-spey and Aberlour.

distillery facts

- 1959
- Allied Distillers Ltd.
- John Black
- The Achvochkie Burn
- 4 wash 4 spirit
- N/A
- Please telephone for an appointment

ages, bottlings, awards

Normally bottled at 10 years and available in the U.K.
Other ages and special bottlings from Gordon & MacPhail

The distillery was built for the Long John Group and is now part of the Allied Distillers portfolio. In 1972 the number of stills was increased from four to eight. Tormore is a golden malt with a well-rounded taste, recommended for after-dinner drinking.

tasting notes

AGE: 10 years 40%

NOSE: A dry aroma with a slightly nutty overtone.

TASTE: Soft on the tongue, a well-defined medium-flavored malt with a hint of honey.

Tullibardine

HIGHLAND

TULLIBARDINE DISTILLERY, BLACKFORD, PERTHSHIRE PH4 1QG
TEL: +44 (0)1764 682252

DETAILS ARE on record of a Tullibardine distillery operating as early as 1798, whose exact site is unknown. This original distillery closed in 1837. Today's Tullibardine was built in 1949 by Delme Evans and C. I. Barratt on the site of a brewery. In 1953 the Tullibardine Distillery Co. Ltd. was acquired by the Glasgow-based whisky brokers, Brodie Hepburn. In 1971 the distillery became part of Invergordon Distillers (Holdings) Ltd, now Whyte & Mackay. In 1973 the distillery was rebuilt and the number of stills increased from two to four. Although the distillery has been mothballed since January 1995, there are sufficient supplies available of this single malt.

distillery facts

- 1949
- The Whyte & Mackay Group Plc
- The Ochil Hills
- 2 wash 2 spirit
- American white oak
- **i** No visitors

tasting notes

AGE: 10 years 40%

NOSE: Light, warm, sweet.

TASTE: Well-rounded malt with hints of fruit and spice with a long, sweet finish.

Yoichi

JAPAN

HOKKAIDO DISTILLERY, KUROKAWA-CHO 7 CHOME-6, YOICHI-CHO,
YOICHI-GUN, HOKKAIDO 046, JAPAN
TEL: +81 (0)135 23 3131 FAX: +81 (0)135 23 2202

WHEN MASATAKA Taketsuru returned from Scotland, after studying whisky distilling at Glasgow University, he looked for a site with ideal conditions for distilling. He found the place to build a whisky distillery on Hokkaido Island at Yoichi. The site is surrounded by mountains on three sides and the ocean on the fourth. Hokkaido is the northernmost Japanese island with cool, clean air and a ready supply of water from underground springs rising through peat bogs.

The distillery was built in 1934 and produces a vibrant copper-colored single malt whisky.

distillery facts

- 1934
- Nikka Whisky Distilling Co. Ltd.
- Hiroshi Hayashi
- Underground springs
- 4 wash 3 spirit
- Mix of sherry, bourbon, refill, and new
- All year round

ages, bottlings, awards
Yoichi is bottled at 12 years and 10,000 bottles are produced annually
The whisky is normally only available in Japan

tasting notes

AGE: 12 years

NOSE: Peaty with a hint of sherry.

TASTE: Full-bodied with a peaty taste and a long finish.

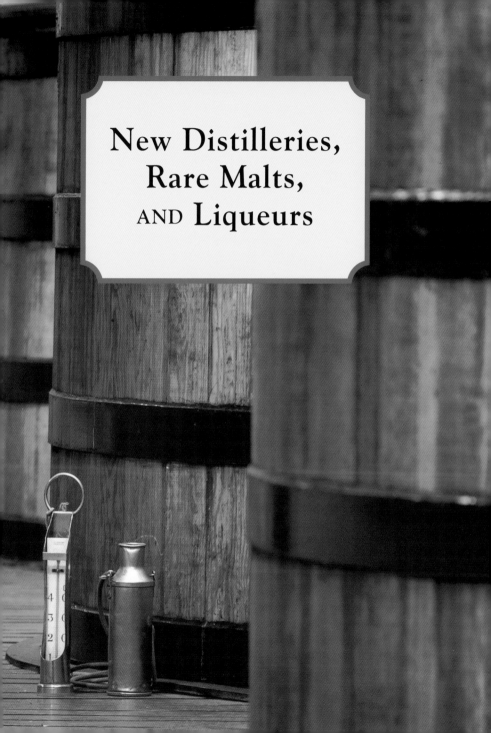

New Distilleries,
Rare Malts,
AND Liqueurs

New Distilleries

There are three new distilleries. Balblair and Old Pulteney are not strictly new, but their new owners have yet to bottle these malts.

Kininvie

SPEYSIDE

KININVIE DISTILLERY, DUFFTOWN, KEITH, BANFFSHIRE AB55 4DH

This distillery was founded in 1990 by William Grant & Sons Ltd., owners of Glenfiddich and Balvenie. The distillery has four spirit and four wash stills and is currently managed by Mr. W. White. With such prestigious owners as William Grant & Sons Ltd. it is sure to be a very special malt.

Old Pulteney

HIGHLAND

PULTENEY DISTILLERY, HUDDART STREET, WICK, CAITHNESS KW1 5BD

TEL: +44 (0)1955 602371 FAX: +44 (0)1993 602279

Inver House Distillers Ltd. purchased Old Pulteney last year, a distillery founded in 1826 by James Henderson, when Wick was a thriving herring fishing port. The distillery was previously owned by Allied Distillers and stocks of Old Pulteney are available at eight and 15 years from Gordon & MacPhail.

Balblair

HIGHLAND

BALBLAIR DISTILLERY, EDDERTON, TAIN, ROSS-SHIRE IV19 1LB

TEL: +44 (0)1862 821273 FAX: +44 (0)1862 821360

Another distillery purchased from Allied Distilleries by Inver House Distillers Ltd. in 1996. Balblair was founded in 1790 and is therefore one of the oldest malt whisky distilleries in Scotland. Inver House are currently reviewing stocks so that they can assess the premium age for a single malt. Stocks of Balblair from Allied Distillers are still available in some specialist retailers and wine merchants.

Rare Malts

Some single malt whiskies have names which linger in the memory, but which cannot be found at your local retailer. This section lists the distilleries which are closed forever, have been mothballed and stocks are low, or are only used in blends. In many instances malt whiskies are only available as special bottlings and are difficult to find.

Alt-a-Bhanie

Part of Seagram Distillers Plc., the distillery was founded in 1975 and the malt is used in the company's blended whiskies only.

Balmenach

A part of United Distillers malt whisky portfolio, Balmenach was mothballed in 1993. Balmenach is a Speyside malt and quite widely available at 12 years 43%.

Banff

This Highland distillery was closed in 1983 and dismantled. A rare malt whisky available from Gordon & MacPhail and Cadenheads.

Braeval

Part of Seagram Distillers Plc., the distillery was founded in 1973 and the malt is used in the company's blended whiskies only.

Coleburn

This Speyside United Distillers distillery closed in 1985 and will not reopen. Some Coleburn 1972 is available from Gordon & MacPhail.

Glen Albyn

This Highland distillery was dismantled in 1986. It is available from Gordon & MacPhail and Cadenheads.

Glenglassaugh

The distillery was mothballed in 1986 and belongs to The Highland Distilleries Company Plc. Glenglassaugh is a Highland malt and Glenglassaugh 1983 is available from Gordon & MacPhail.

Glen Mohr

The distillery was closed in 1983 and dismantled in 1986. Stocks of Glen Mohr are available from Gordon & MacPhail.

Glen Mohr

The distillery was closed in 1983 and dismantled in 1986. Stocks of Glen Mohr are available from Gordon & MacPhail.

Glen Scotia

This Campbeltown distillery, which belongs to Loch Lomond Distillery Co. Ltd. was mothballed in 1994. Stocks are available from the distillers of Glen Scotia 14 years at 40% for home and 43% for export sales.

Glenugie

Closed in 1983 and only available from Cadenheads.

Glenury Royal

Part of the United Distillers malt whisky portfolio, the distillery closed in 1985 and will not reopen. Available at 12 years 40%.

Inverleven

Owned by Allied Distillers, Inverleven forms part of their Dumbarton grain distillery complex which sits on the mouth of the River Leven as it flows into the River Clyde. Inverleven was mothballed in 1989. Stocks of 1984 are available from Gordon & MacPhail.

Littlemill

This lowland distillery which belongs to Loch Lomond Distillery Co. Ltd was mothballed in 1992. Stocks of eight years 40% and 43% for export are quite widely available.

Lochside

This Allied Distillers distillery, which is located in Montrose, was closed in 1991. Stocks are available of Lochside ten years from the distillery office.

Millburn

This distillery closed in 1985 and was dismantled in 1988. Some stocks of this highland malt are available from Gordon & MacPhail.

Pittyvaich

This United Distillers distillery was mothballed in 1993, although some Pittyvaich is available at 12 years 43%.

Port Ellen

This Islay distillery, owned by United Distillers closed in 1983. Stocks of Port Ellen 1979 are available from Gordon & MacPhail. Port Ellen was the first distillery to export direct to America in the 1840s.

St. Magdalene

This lowland distillery closed in 1983 and has now been converted into housing. St. Magdalene 1966 is still available from Gordon & MacPhail.

Spey Royal

International Distillers & Vintners use this malt from their Glen Spey Distillery in Rothes in their blended whiskies.

Tamnavulin

Part of The Whyte & Mackay Group PLC, Tamnavulin was mothballed in 1995. Some bottles are still available of ten years old 40% Tamnavulin.

At the bar at the Athenaeum Hotel.

Malt Liqueurs

There are a surprisingly large number of malt liqueurs, both Scottish and Irish.

Malt Liqueurs from Scotland

Drambuie

Available in bottles and miniatures at an alcoholic strength of 40%. Drambuie is marketed as Bonnie Prince Charlie's liqueur. The recipe is reputed to have been given to Captain John Mackinnon for his loyalty at the Battle of Culloden in 1746. Drambuie is a sweet liqueur with honey and fruit flavors.

Dunkeld Atholl Brose

Atholl Brose is a traditional recipe and is a mixture of oatmeal, honey, water, and whisky. Dunkeld Atholl Brose liqueur is produced by Gordon & MacPhail and sells at 12 years old with an alcoholic strength of 35%.

Glayva

Glayva liqueur is marketed by The Whyte & Mackay Group. Glayva means "very good" in Gaelic. A sweet liqueur recommended for after dinner, Glayva has a smooth texture and a hint of citrus.

The Glenturret
Original Single Malt Liqueur

The Glenturret Distillery is one of the oldest malt whisky distilleries, founded in 1775. The fine single malt produced at this distillery is used in The Glenturret Malt Liqueur, which is blended with herbs. This is a well flavored, smooth liqueur, which is good on its own and also mixes well with soda or lemonade to make a long refreshing drink.

Heather Cream Liqueur

Heather Cream is part of Inver House Distillers. This company owns An Cnoc, Speyburn, Pulteney, and Balblair distilleries. Heather Cream is a mixture of cream and malt whisky and is a sweet, smooth liqueur.

Stag's Breath Liqueur

This rather unprepossessing name is taken from Sir Compton Mackenzie's book *Whisky Galore* based on the sinking of the *S.S. Politician*, when whisky was lost overboard. Mackenzie named one of the lost whiskies Stag's Breath. Stag's Breath is a mixture of Speyside whisky and honey.

Wallace
Single Malt Liqueur

Wallace Single Malt Liqueur is marketed by Burn Stewart Distillers and is made from Deanston single malt whisky and a mixture of berries and herbs to produce a warm satisfying after-dinner liqueur or mixer with soda, fruit juices, and ice.

Malt Liqueurs from Eire

Baileys

Baileys Original Irish Cream is produced in Dublin by R. & A. Bailey & Co. and is a market leader. Baileys is a sweet liqueur with honey, chocolate and whiskey with an alcoholic strength of 17%. Baileys Light, a lower-calorie and lower-fat version, is widely sold in the United States.

Carolans

Carolans Irish Cream liqueur was first marketed in 1978 and is a very popular after-dinner drink. Carolans also market Carolans Irish Coffee Cream which sells mainly in the United States.

Eblana

Eblana is a new liqueur from Cooley Distillery with an alcoholic strength of 40% with a full sweet flavor.

Emmets

Another cream liqueur from the same stable as Baileys. The liqueur is named after Robert Emmet, an Irish hero who was executed in 1803 for his part in a revolt against British rule.

Irish Mist

Irish Mist is produced by the same company as Carolans and has been marketed since the early 1950s. Irish Mist is a flavorful liqueur with herbs, honey, and whiskey.

Sheridans

Sheridans is bottled in distinctive bottles at 17% in its vanilla cream version and 19.5% in its coffee chocolate version. Another stablemate in the Baileys portfolio, Sheridans liqueurs are rich and warming.

Appendix

Glossary of Terms

Scottish words for geographical details

Ben	Hill or mountain
Burn	A stream
Loch	A lake sometimes with a river flowing in and out each end, often surrounded by mountains
Paps	Mountains, particularly associated with the Paps of Jura, three high mountains on the Island of Jura on the West Coast of Scotland
Rill	A small stream or brook

Distilling terminology

Barrels or casks — Distilleries use various types of casks in which to mature their whisky:

Cask type	Approximate content in U.S. gallons
Butt	100
Hogshead	60–70
American Barrel	35–42
Quarter	27–34

Cask Strength Whisky — This is whisky sold at the strength taken from the cask. Cask strength is normally 60% alcoholic volume or 120° proof.

Coal-fired and-Steam-driven Stills — Stills are heated at the base so that the liquid inside is heated and the alcoholic vapor released. In Scotland early distilleries were heated directly using coal or by steam produced by steam driven engines.

Coffey stills — Aeneas Coffey invented the Coffey still in 1831. The principle behind a Coffey still is simple; continuous production of spirit without having to empty and refill pot stills. Grain whisky is produced in a Coffey still and typically contains about 25%

malted barley plus unmalted barley and maize. Fermentation is as described in the introduction to this book, but produced in a continuous flow and pumped into the top of a large column, or rectifier, through which it flows in a zigzag pattern. The wash is then pumped to the top of a nearby analyzer column which contains a number of perforated plates. The liquid flows through the plates and at the same time steam is forced upwards from the bottom. The steam collects the alcohol as it rises through the column and leaves the spent lees at the bottom. The alcohol laden steam is then taken to the bottom of the rectifier column and the spirit condenses as it flows up the column. As the spirit reaches the top of the column it is taken off for maturation. Coffey stills are between 40 and 50 feet high. Coffey stills are not used for the production of single malt whiskies.

Cooper	A craftsman skilled in making and repairing wooden barrels and casks.
Customs & Excise Officer's House	Once whisky distilleries were licensed, HM Customs & Excise started to control the production of whisky. To ensure that everything was carried out legitimately Excise Officers lived on the site and a house was often built for them by the distillery companies. This practice has now stopped and it is the distillery manager's responsibility to control whisky stocks.
Draff	The solids which remain at the base of the mash tun and are removed and used as cattle feed.
Feints	After the pure spirit has been taken off from the spirit still the condensed vapor weakens and is no longer pure. This weakened spirit is known as feints and is discarded.
Foreshots	The first liquid produced in the spirit still, as the steam condenses. Foreshots turn cloudy when water is added as the spirit is still impure and is discarded.
Grist	The dried malt is ground finely and is then known as grist.
Lyne arms	The top of a pot still bends to form an arm through which the

spirit passes into the spirit still. The shape of these arms varies and distillery managers believe that the different arms contribute to the final characteristics of the whisky. One type is known as a lyne arm.

Malted barley
Germinated malt is known as malted barley when the enzymes in the barley have been released, which give barley its malty taste.

Malting floor
In a traditional distillery barley is soaked in water for two or three days and then spread on a stone malting floor until germination takes place.

Malt kiln
A traditional method of drying malted barley is using smoke from a kiln usually burning peat. The smoke filters through a fine mesh to the barley above.

Mashing
Grist is then mixed with hot water in a mash tun.

Mash tuns
Large circular vessels often made from copper, with a lid. Mechanical rakes move around inside the mash tun to ensure that the barley is mixed with the boiling water. The base of the tun has filter panels, which allow the liquid to drain off and the solids remain inside the mash tun.

Mothballed
Some distilleries have been closed down for a while yet could reopen any time. The Scotch whisky industry talks of such distilleries as being mothballed; everything is kept in pristine condition until the day comes for production to start again.

Saladin Maltings
A mechanically controlled method of germinating barley. The barley is placed in large rectangular boxes. Air is blown up through the grain at controlled temperatures and the grain is turned mechanically.

Single Cask Whisky
This is whisky from just one cask or barrel, bottled usually in a numbered limited edition. This single malt is either bottled at cask strength or diluted and bottled at 40% or 80° proof.

Spirit stills
These stills are used for the second distillation and spirit is collected from these to be stored into barrels.

Spent lees
Waste material produced during distillation.

Stills
These are traditionally made from copper and in the case of malt

whisky distillation are also known as pot stills. Pot stills produce whisky in batches, not continuously.

Triple distillation Triple-distilled single malt whisky is produced by passing the spirit through the spirit still twice.

Wash The liquid drawn off from the mash tun is commonly called wort, but sometimes described as wash.

Washbacks These are large vessels, usually wooden, which can hold from 3,000 to 15,000 gallons. The liquid (wort) from the mash tun is pumped into the washbacks and yeast added to convert the wort into alcohol.

Wash stills The fermented liquid from the washbacks is pumped into these stills for the first distillation which separates out the alcohol.

Wort The liquid drawn off from the mash tun is known as wort. The solid material resting at the bottom of the mash tun is used for cattle feed.

Other points of interest

Beeching 1967 Dr. Thomas Beeching was responsible for the rationalization of the railways in the United Kingdom in 1967. Rationalization meant that many smaller branch lines were closed and isolated communities lost their only means of local transport.

Whisky Recipes

Whisky is usually served as an aperitif or after-dinner drink. However, many people enjoy whisky as part of a cocktail. Whisky can also be served throughout the meal, particularly when celebrating St. Andrew's night (the patron saint of Scotland) or at a traditional Burns' Supper, held on January 25. Robert Burns (1759–96) was a great Scottish poet and wrote about whisky in several of his poems, notably the New Year's song "Auld Lang Syne" where "the cup o'kindness" refers to a glass of Scotch whisky. At such Scottish celebrations haggis is served and it is customary to pour a glass of whisky over the haggis. (Haggis is a large round sausage made with onion, spices, oatmeal, and lamb.)

COCKTAILS

whisky collins

1 measure of Scotch whisky
1 tsp. sugar syrup
Club soda
Juice of ½ lemon
Angostura Bitters

In a tall glass put some ice and the lemon juice. Add the sugar syrup and whisky. Fill with club soda and a few drops of Angostura. Stir and serve with a slice of lemon.

john milroy's hot toddy

1 tsp. honey
1 measure malt whisky
1 measure green ginger wine
Juice of ½ lemon
1 clove
Boiling water

Dissolve the honey in a little boiling water. Add the clove, lemon juice, malt whisky, and ginger wine. Stir and top up with 3 measures of boiling water.

bobby burns

½ measuring cup of Scotch whisky

¼ cup of dry vermouth

¼ cup of sweet vermouth

A dash of Benedictine

Stir all the ingredients in a glass filled with crushed ice. Serve with a twist of lemon peel.

scotch old fashioned

I measure of Scotch whisky

3 dashes of Angostura Bitters

Small sugar lump

Place the sugar in a glass. Add the Angostura and a little water to dissolve the sugar. Stir in the whisky and some ice. If liked add a slice of orange and a cherry.

COOKING WITH WHISKY

Cooking soups, salads, steaks, and desserts with whisky may seem surprising, but the results are particularly good. For hors d'oeuvres and main courses blended whisky will suffice, but for dessert a single malt is well worth the extra expense. All recipes given below are for four people.

avocado cocktail

2 avocados peeled and cut into
    small pieces

Salad dressing made with olive oil,
    blended whisky, lemon juice, pinch of
    sugar, pepper, and salt.

Lettuce

Place lettuce on each plate, scatter with the avocado and sprinkle with the dressing.

marinated mushrooms

6 oz. button mushrooms, sliced finely

3 tbsp. whisky

3 tbsp. oil

½ juice of a lemon

A pinch of sugar

A few crushed coriander seeds

Pepper, salt, and mixed herbs

Simply put all the ingredients in a bowl, mix thoroughly, cover, and leave for about 1½ hours, stirring occasionally.

steak & scotch

Prepare your steak as usual in a skillet. Remove the steak and place in the oven to keep warm. Return the skillet to the stove top and add a measure of whisky to the juices in the pan. When the mixture has reduced slightly pour over the steak and it is ready to serve.

whisky creams

2 tbsp. of malt whisky

8 fl. oz. milk

2 fl. oz. heavy cream

4 egg yolks

3 tbsp. of sieved orange marmalade

A pinch of nutmeg

Heat the milk and cream in a pan. Beat the egg yolks, whisky, spice, and marmalade together and pour onto the milk. Carefully heat the mixture in a bowl over a pan of hot water until it is thick and creamy. Pour into individual dishes and leave to cool.

smoked salmon pasta

Whisky goes particularly well with lobster, scallops, and salmon.

8 oz. pasta (suggest penne or rigatoni)

4 oz. smoked salmon sliced into thin strips

6 fl. oz. heavy cream

2 garlic cloves

1 oz. butter

1 medium onion cut finely

Salt, pepper

Parmesan cheese

1 measure of whisky

Heat the heavy cream in a saucepan with the garlic and boil until the garlic cloves are soft. Remove the garlic and set the saucepan aside. Fry the onion in the butter until golden, but not brown, and pour into the flavored cream. Stir the cream mixture until the cream begins to thicken, then add salt and pepper, and the glass of water.

Meanwhile cook the pasta in a pan of boiling salted water, drain thoroughly. Add the sauce to the pasta, stir in the smoked salmon and serve piping hot with parmesan cheese on the side.

Enjoying Single Malt Whisky

First time drinkers of single malts may not wish to invest in a full bottle straight away. Whiskies are also available as miniatures and this is a good way to start. Many restaurants and hotels throughout the world offer a good selection of single malt whiskies for the whisky drinker to taste and enjoy.

HOTELS, RESTAURANTS, AND PUBS IN THE UNITED KINGDOM

Most major hotels in the United Kingdom have good whisky collections.

The Athenaeum Hotel in London's Piccadilly (see p. 237) has a very special Malt Whisky Bar with more than 70 different malts. Visitors can obtain a "passport" from the bar listing the whiskies available, which is stamped each time a different malt is chosen. Visitors to the Nobody Inn in Doddiscombleigh in Devon will also find a wide choice of single malts.

In Scotland there are numerous hotels and inns selling single malt whisky. Gleneagles and Turnberry Hotel offer guests and golfers alike a wide choice of single malts. At The Borestone Bar in Stirling, there are 1,000 different malts behind the bar.

HOTELS, RESTAURANTS, AND BARS IN THE UNITED STATES

Most leading hotels in the United States offer a good selection of Scotch whiskies. The increasing popularity of cigar smoking has given rise to a number of cigar bars that offer customers a selection of single malts. These include Beekman Bar and Books and Club Macanudo in New York, Three of Clubs and Bar Marmont in Los Angeles, Berlin Bar in Miami Beach, Occidental Grill in San Francisco, and Fumatore Cigar Bar & Club in Chicago.

A selection of restaurants is listed on p. 254

Buying Single Malt Whisky

Where to buy whisky in the United Kingdom

Most High Street liquor stores stock a selection of single malts. These will vary from store to store and according to the time of year. More malt whiskies are available at Christmas. Oddbins probably carries the widest range of single malt whiskies and often has special offers, which are well worth looking out for.

There are some distilleries which will never produce malt whisky again, for example St. Magdalen, which is now a housing development. There are others, which are uneconomical to reinstate. However, before they were closed down, their stocks of fine malt whiskies were put into casks and are slowly being made available to the whisky connoisseur. These specialist bottlings are included in the Directory under Rare Malts. In addition there are special bottlings perhaps of just one cask, sometimes at an age not normally available. These are to be found from a specialist retailer or through societies. The listing below is by no means a definitive one, but gives an indication of where the diligent whisky drinker may find hidden treasures.

Gordon & MacPhail is a family business and was founded in 1895. They stock an extremely wide range of whiskies. These include the Connoisseurs Choice brand. Many of the whiskies represented in this brand are not normally available as single malts, as the entire production of the distillery concerned might be used for blending or may no longer exist. Gordon & MacPhail has recently revived the distillery at Benromach, which has remained silent for many years. The company also markets a range of vatted malts under the Pride of the Regions label. From their shop just below Edinburgh Castle the company markets a wide range of cask strength bottled malts (some 57 percent), many of which are at ages found nowhere else. In addition the company has a range of standard bottlings, which are normally at 46%.

John Milroy (see p.253) has recently launched his own independent bottling and whisky brokerage company, and his products will certainly be worth seeking out.

The Scotch Malt Whisky Society (see p.254) society was founded in 1983 and offers its members a choice of malts taken straight from the cask. These are of a considerably greater strength than standard bottlings and as each cask is different, the range of malts is considerable. The society, which has a prestigious base in Leith, the old port of Edinburgh, offers its members the use of a lounge bar, tasting room, and other facilities. The society also organizes tastings throughout the United Kingdom. There are Scotch malt whisky societies in France, Switzerland, The Netherlands, Japan, and the United States. Details of these can be obtained from the Edinburgh office.

WHERE TO BUY WHISKY IN THE UNITED STATES

Many Americans purchase their whisky from specialist retailers in the United Kingdom, however there are many companies in the United States selling a wide range of single malts and blended whiskies. (see p.254)

The Scotch Malt Whisky Society, (listed on p.254), publishes a regular newsletter together with bottling lists which describe those whiskies which are currently available and from which supplies can be ordered. Orders can be placed by mail using the order form, by telephone, or by fax. Members' orders will be fulfilled by express delivery from a selected liquor purveyor from within your state. The membership fee includes the purchase price of one 750ml bottle of an extremely rare and unique malt never to be available again. Thereafter subscriptions are renewable for a modest fee on the anniversary of joining.

Investing in Whisky

At the present moment, the subject of investing in whisky is achieving a great deal of publicity. Companies outside the Scotch whisky industry are suggesting that buying whisky is a lucrative investment. The Scotch Whisky Association has published a leaflet "Personal Investment in Scotch Whisky in Cask" and warns investors that "the industry does not work in a way which is conducive to investment. It is unregulated and there is no whisky exchange on which to trade." The author's advice is that if you wish to buy a barrel of whisky, then do so, but only for pleasure, do not assume you will make a lot of money when the whisky matures. Purchasers of a cask of whisky should also bear in mind that when the whisky is bottled it will become eligible for excise duty (tax) at the current rates, not the rate of duty applicable when the cask was purchased.

WHISKY AUCTIONS

Christie's in Scotland handle specialist whisky auctions from time to time. Prices for individual bottles of single malts can be very high. For example at an auction of Thursday, May 9, 1996, a bottle of The Glenlivet Jubilee Reserve, bottle number 506—25 year old, 75° proof, packed in a wooden presentation box sold for $510.

Whisky auctions provide the connoisseur with the opportunity of buying a very special malt and it is true that such special bottles will increase in value, albeit slowly.

LIMITED EDITION PACKAGING, SPECIAL BOTTLINGS, AND MINIATURES

For those wishing to form a malt whisky collection it is comparatively easy to find special bottlings and limited edition packaging. And who knows, your special bottling or packaging could be worth something in the future. Do not drink the whisky and keep the packaging as clean as possible. Collecting miniatures is fascinating as many malts are bottled as souvenirs and the number of different bottlings for each malt can be considerable.

Useful addresses

It would be an impossible task to give details of all the distributors and societies associated with single malt whisky, but this list offers a good international selection in addition to the major companies in the United Kingdom. Also listed are places to go to buy and drink single malts.

Distributors

AUSTRALIA

Allied Domecq Spirits &
Wine PTY Ltd.
Suite 704 7th Floor
7 Help Street
Chatswood
New South Wales 2067
Tel: +612 9411 7077
Fax: +612 9413 2902

Remy Australie Ltd.
484 Victoria Road
Gladesville
New South Wales 2111
Tel: +612 9816 5000
Fax: +612 9817 3170

FRANCE

Baron Phillippe de
Rothschild
France Distribution
64 Bis, Rue La Boetie
75008, Paris
Tel: +33 1 44 132 020
Fax: +33 1 42 560 101

Remy Distribution
France
126 Rue Jules Guesde
9230 Levallois-Perret
Tel: +33 1 4968 4968
Fax: +33 1 4370 4968

GERMANY

Herman Joerss GmbH
Sohnleinstrasse 8
6200 Wiesbaden
Tel: +49 611 25002
Fax: +49 611 250420

JAPAN

Berry Bros & Rudd Ltd.
Shinwa Building
6F, 2-4 Nishi Shinjuku
3-chome,Shinjuku-ku
Tokyo 106
Tel: not available

Nikka Whisky Distilling
Co., Ltd.
4–31 Minami Aoyanma
5-chome, Minato-ku 105
Tel: +81 3 3498 0331
Fax: +81 3 3498 2030

Pernod Ricard Japan
K.K. 3rd Floor
Shinagawa NSS Bldg.
13–1 Toranoman
5-chome,
Minato-ku 105
Tel: +81 3 3359 2266
Fax: +81 3 3359 2224

Remy Japon K.K.
Mori Bldg 13/1
Toranomon 5-chome
Minato-ku, Tokyo
Tel: +81 3 5401 6272
Fax: +81 3 3434 8425

Suntory Limited Liquor
Division 1-2-3
Motoakasaka Minato-ku
Tokyo 107
Tel: +81 3 3470 1183
Fax: +81 3 3470 1330

United Distillers Ltd.
Sumitomo Gotanda
Bldg.
1–1 Nishi Gotanda
7-Chome
Shinagawa-ku 141
Tel: + 81 3 3491 3011
Fax: + 81 3 3492 1830

UNITED KINGDOM

Allied Distillers Ltd.
Kilmalid
Dumbarton
G82 2SD
Tel: +44 1389 765111
Fax: +44 1389 763874

Ben Nevis
Distillery Ltd.
Lochy Bridge
Fort William
PH33 6TJ
Tel: +44 1397 702476
Fax: +44 1397 702768

Berry Bros & Rudd Ltd.
3 St. James Street
London SW14 1EG
Tel: +44 (0)20 7396 9666

Burn Stewart
Distillers Plc.
8 Milton Road
College Milton North
East Kilbride
G74 5BU
Tel: +44 1355 260999
Fax: +44 1355 264355

Glenmorangie Plc.
Macdonald House
18 Westerton Road
Broxburn
West Lothian
EH52 5AQ
Tel: +44 1506 852929
Fax: +44 1506 855055

Gordon & Macphail
George House
Boroughbriggs Road
Elgin Moray
IV30 1JY
Tel: +44 1343 545111
Fax: +44 1343 540155

Inver House
Distillers Ltd.
Airdie
Lanarkshire
ML6 8PL
Tel: +44 1236 769377
Fax: +44 1236 769781

Justerini & Brooks Ltd.
8 Henrietta Place
London W1M 9AG
Tel: +44 (0)20 7518 5400
Fax: +44 (0)20 7518 4651

Matthew Gloag &
Sons Ltd.
West Kinfauns
Perth PH2 7XZ
Tel: +44 1738 440000
Fax: +44 1378 618167

Morrison Bowmore
Distillers Ltd.
Springburn Road
Carlisle Street
Glasgow, G21 1EQ
Tel: +44 141 558 9011

United Distillers
Distillers House
33 Ellersly Road
Edinburgh, EH12 6JW
Tel: +44 131 337 7373
Fax: +44 131 337 0163

Whyte & Mackay
Dalmore House
310 St. Vincent Street
Glasgow, G2 5RG
Tel: +44 141 248 5771
Fax: +44 141 221 1993

William Grant & Sons
Independence House
84 Lower Mortlake Rd
Richmond, Surrey
TW9 2HS
Tel: +44 (0)20 8332 1188
Fax: +44 (0)20 8332 1695

UNITED STATES

Allied Domecq Spirits &
Wine Limited
300 Town Center
Suite 3200
Southfield
Michigan 48075
Tel: +1 810 539 3218

Palace Brands Company
450 Columbus Boulevard
PO Box
778 Hartford
CT 06142-0778
Tel: +1 860 702 4421
Fax: +1 860 702 4489

Remy Amerique
1350 Avenue of the
Americas 7th Floor.
New York, NY 10019
Tel: +1 212 399 9494
Fax: +1 212 399 2461

United Distillers
6 Landmark Square
Stamford, CT 06901
Tel: +1 3203 359 7100
Fax: +1 203 359 7196

Societies and Associations

EUROPE

The Scotch Malt Whisky Society B.V.
Vijhuizenberg 103,
PB 1812
4700 BV Roosendaal
The Netherlands
Tel: +31 1650 33134
Fax: +31 1650 40067

The Scotch Malt Whisky Society France
171 Rue de Charenton,
BP 145
75562 Paris CEDEX 12
Tel: +33 1 44 75 5353
Fax: +33 1 44 75 5354

The Scotch Malt Whisky Society Switzerland
Kraan and Richards Imports
Gartenstrasse 99, CH-4052 Basel
Tel: +41 61 271 5460
Fax: +41 61 272 4123

JAPAN

The Scotch Malt Whisky Society Japan
15/32 Nakanocho
2-chome
Mikakojimaku
Osaka 534
Tel: + 81 6 351 9145
Fax: +81 6 351 9198

UNITED KINGDOM

The Scotch Whisky Association
14 Cork Street
London W1S 3NS
Tel: +44 (0)20 7629 4384
Fax: +44 (0)20 7493 1398

The Scotch Malt Whisky Society
The Vaults
87 Giles Street
Leith
Edinburgh EH6 6BZ
Tel: +44 131 554 3451
Fax: +44 131 555 6588

The Scotch Malt Whisky Society
9838 West Sample Road
Coral Springs
Florida 33065
Tel: +1 954 752 7990
Fax: +1 954 752 8552

Stores and Bars

UNITED KINGDOM

Cadenheads Whisky Shop
172 Canongate
Edinburgh
EH8 5BH
Tel: +44 (0)131 556 5864

Milroy's of Soho
3 Greek Street
London W1V 6NX
Tel: +44 (0)20 7437 0893

UNITED STATES

Keen's Chop House
72 West 36th Street
New York

The Post House
28 East 63rd Street
New York

Tavern on the Green
Central Park at 67th Street

Morrell & Company
535 Madison Avenue
New York

Park Avenue Liquor Shop
292 Madison Avenue
New York

Sam's Wine Warehouse
1720 North Macey Street
Chicago

Sherry-Lehmann
679 Madison Avenue
New York

Picture Credits:

The publisher would like to thank all of the individual distilleries and their owning companies for contributing illustrative material to support their entries in this book. Additional picture credits go to the following:
The Glenturret Distillery Co. p1; Morrison Bowmore Distillers Ltd., pp.9, 18, 19; Alaister Skakles; pp.1,11; Matthew Gloag & Son, pp. 12, 17, 20, 22, 25, 27; William Grant & Son, pp. 13, 26 and black and white illustrations in Part I; Helen Arthur, pp. 20(t), 21; Allied Distillers Ltd. p.23; Life File Photographic Agency, pp. 24, 28, 29, 31, 32, 33, 34, 35, 36, 38, 78, 91, 103,117, 149, 180, 186, 197, 255; Edinburgh Crystal p.48 (l).

While every effort has been made to ensure that all credits are listed, the Publisher apologizes for any omissions.